Electroactivity in Polymeric Materials

T0134471

Lenore Rasmussen
Editor

Electroactivity in Polymeric Materials

Lenore Rasmussen
Ras Labs, LLC, Intelligent Materials for Prosthetics and Automation
Plasma Surface Modification Experiment
US Department of Energy's Princeton
Plasma Physics Laboratory at Princeton University
Room L-127, 100 Stellarator Road
Princeton
NJ 08543
USA

ISBN 978-1-4899-8551-4 ISBN 978-1-4614-0878-9 (eBook)
DOI 10.1007/978-1-4614-0878-9
Springer New York Heidelberg Dordrecht London

© Springer Science+Business Media New York 2012
Softcover reprint of the hardcover 1st edition 2012
This work is subject to copyright. All rights are reserved by the Publisher, whether the whole or part of the material is concerned, specifically the rights of translation, reprinting, reuse of illustrations, recitation, broadcasting, reproduction on microfilms or in any other physical way, and transmission or information storage and retrieval, electronic adaptation, computer software, or by similar or dissimilar methodology now known or hereafter developed. Exempted from this legal reservation are brief excerpts in connection with reviews or scholarly analysis or material supplied specifically for the purpose of being entered and executed on a computer system, for exclusive use by the purchaser of the work. Duplication of this publication or parts thereof is permitted only under the provisions of the Copyright Law of the Publisher's location, in its current version, and permission for use must always be obtained from Springer. Permissions for use may be obtained through RightsLink at the Copyright Clearance Center. Violations are liable to prosecution under the respective Copyright Law.
The use of general descriptive names, registered names, trademarks, service marks, etc. in this publication does not imply , even in the absence of a specific statement, that such names are exempt from the relevant protective laws and regulations and therefore free for general use.
While the advice and information in this book are believed to be true and accurate at the date of publication, neither the authors nor the editors nor the publisher can accept any legal responsibility for any errors or omissions that may be made. The publisher makes no warranty, express or implied, with respect to the material contained herein.

Printed on acid-free paper

Springer is part of Springer Science+Business Media (www.springer.com)

Preface

The thoughts of making a theoretical textbook style book in electroactive materials began when I started getting more students in the laboratory. When new students join the project, I invariably give them a stack of journal articles to wade through, so began thinking how helpful it would be to give them a concise book to start with. This book is designed as a starting point, particularly to get a handle on "how does electroactive movement happen?"

The goal of this book was to capture the theory—how electroactivity works—balanced with applications—how can electroactivity be used, drawing inspiration from our manmade mechanical world and the natural world around us. This book takes a small step at capturing the fascinating field of electroactive materials and actuators. I'm already putting thoughts together on how to make the second volume even more informative while retaining concision and clarity.

The Artificial Muscle Project draws people from a variety of disciplines. Indeed, the field of electroactivity is extremely interdisciplinary. Case in point: for my first patent in this area, the examiner from the USPTO called me. Evidently, they had several meetings trying to decide which patent classification code it came under—chemistry or electrical engineering? So they resolved the matter by putting the question to me. We chatted and I agreed that it was on the fence between the two areas, but since my background was stronger in chemistry, the main classification was placed in class 523/113, synthetic resins, subclass composition suitable for use as tissue or body member replacement, restorative, or implant.

Fundamentally, the thorough understanding of electroactivity is important because of the ability of bending, contraction, and expansion to produce smooth, controllable, life-like biomimetic motion. By combining electroactive materials with fuel cells and other technologies, electroactive actuation and its reciprocal action can also provide for extremely energy efficient motion, energy generation, and energy harvesting. Electroactivity offers new ways of thinking about and configuring devises, machines, implants, and surfaces, for futuristic mobility by land, air, and sea.

Acknowledgments

The editor would like to thank everyone who contributed to this effort, both for advancing the field of electroactivity as well as those who helped with the preparation of this book. The editor would like to thank the other co-authors: Prof. Iain Anderson and his group including Todd Gisby and Ben O'Brien; Prof. Mohsen Shahinpoor and his group including Yousef Bahramzadeh; Prof. Qibing Pei and his group including Paul Brochu; Dr. Roy Kornbluh, Ron Pelrine, Harsha Prahad, Annjoe Wong-Foy, Brian McCoy, Susan Kim, Joseph Eckerle, and Tom Low of SRI International; and others who have made so many strides in the field of electroactivity: Dr. Yoseph Bar-Cohen, Profs. Yoshihito Osada and Jian Ping Gong, Prof. Toyoichi Tanaka, Prof. John Madden, Prof. Elizabeth Smela, Prof. Federico Carpi, Prof. Giovanni Pioggia, Prof. Danilo de Rossi, Prof. Cynthia Breazeal, Dr. Emilio Calius, Prof. Selahattin Ozcelik, Prof. Alexie Khokhlov, Prof. Donald Leo, Prof. Timothy E. Long, Prof. Roger Moore, Prof. Qiming Zhang, and many, many others. Together, all of us collectively, are making science fiction a reality.

Paramount to the success of this book was the United States Department of Energy's Princeton Plasma Physics Laboratory at Princeton University. I would like to personally thank Lewis Meixler, Charles Gentile, George Ascoine, Yevgeny Raitses, Eliot Feibush, Philip Efthimion, Adam Cohen, Stewart Prager, Anthony DeMeo, Kitta McPherson, Patricia Wieser, Jim Taylor, Stephan Jurczynski, Carl Tilson, Sue Hill, Gary D'Amico, William Zimmer, John Trafalski, and many others for their support, encouragement, expertise, and state-of-the-art scientific capabilities. I would also like to thank the professors and their laboratories at Princeton University that have helped so much with this endeavor: Prof. Robert Cava and Dr. Anthony Williams for help in the solid state laboratory; Prof. Steven L. Bernasek and Dr. Esta Abelev for their help with X-ray photoelectron spectroscopy; and Jane Woodall and Prof. Nan Yao for their help with scanning electron microscopy at the Image and Analysis Center. I would like to thank Frank Cozzarelli, Jr., for his patent expertise and insights. A special heartfelt thanks goes to Thomas Brown of the Federal Laboratory Consortium.

I would like to thank all the interns and continuing education teachers who have contributed to the Artificial Muscle Project over the years: Alice Kirk, Erich Schramm, Carl J. Erickson, David Schramm, Kelsey Pagdon, Dan Pearlman, Kevin Mulally, Sarah Newbury, Aparna Panja, Victoria Jones, and my two sons, who had no choice but to dragged into the business, Paul and Lars Rasmussen. Your laboratory technique, persistence, inquiring minds, and questions that made me think and re-think, made this project a success. I would also like to thank my youngest son, Carl Rasmussen, for his curiosity in the greater world around us. Many heartfelt thanks to Barbara Jones for her proofing abilities, translation abilities, and wording suggestions.

I would like to thank Dean Kathryn E. Uhrich of Rutgers University and Prof. James E. McGrath, Prof. Garth L. Wilkes, and Prof. Eugene M. Gregory of Virginia Tech, for all their help with my education, critical thinking, laboratory technique, and ability to synthetically tailor materials, all of which has served so well in academia and industry. A special thanks goes to Dan and Judi McGuire for providing a home away from home while completing my education.

Last but certainly not least, I would like to thank the rest of my family for all their love, support and encouragement in this endeavor: my husband Henrik T. Rasmussen for his love and devotion, my mother Winola H. Carman for her continual encouragement, my brothers and sister-in-law Paul Carman, Nathan and Sally Carman, my father R. Wayne Carman, and my beautiful family through marriage, Jørgen and Ingrid Rasmussen, Tom and Linda Rasmussen, and Morten and Carolin Rasmussen. Desperation may be the mother of invention, but encouragement is the foundation of creativity.

Contents

Chapter 1
Dielectric Elastomers for Actuators and Artificial Muscles

Paul Brochu and Qibing Pei

Abstract A number of electroactive polymers have been explored for their use as artificial muscles. Among these, dielectric elastomers appear to provide the best combination of properties for true muscle-like actuation. Dielectric elastomers behave as compliant capacitors, expanding in area and shrinking in thickness when a voltage is applied. Materials combining very high energy densities, strains and efficiencies have been known for some time. To date, however, the widespread adoption of dielectric elastomers has been hindered by premature breakdown and the requirement for high voltages and bulky support frames. Recent advances seem poised to remove these restrictions and allow for the production of highly reliable, high-performance transducers for artificial muscle applications.

Keywords Dielectric elastomer · Electroactive polymer · Bistable electroactive polymers · Actuator · Transducer · Artificial muscle · DE · EAP · BSEP

1.1 Introduction

The ability to mimic the muscles in our own human bodies, both for the advancement in our well-being and for our amusement, has been a topic of great interest for some time. Natural muscle has a number of properties that make it difficult to match in terms of performance. The energy density of muscle is on the order of 150 J kg^{-1} and can peak at around 300 J kg^{-1} [1], while displacements

P. Brochu · Q. Pei (✉)
Department of Materials Science and Engineering, The Henry Samueli School of Engineering, University of California, 420 Westwood Plaza, Los Angeles, CA 90095-1595, USA
e-mail: qpei@seas.ucla.edu

L. Rasmussen (ed.), *Electroactivity in Polymeric Materials*,
DOI: 10.1007/978-1-4614-0878-9_1,
© Springer Science+Business Media New York 2012

are relatively large with typical strains ranging from 20 to 40% and peaking at 100% [2–4]. By these measures alone, electromagnetic (EM) motors and combustion engines should be able to match or exceed the performance of natural muscle [5]. However, as it is made obvious by current leading-edge robots (e.g., Honda's Asimo) [6], the real world performance of conventional-actuator-based robotics is limited [2, 7]. The shortcoming lies on several fronts. First is the power supply: natural muscle relies on chemical energy that is supplied to living organisms through the ingestion of food, while EM motors rely on heavy battery pack and capacitor banks that must be recharged frequently. These large power sources contribute to the overall mass of the robotic device and reduce the effective energy density as well as limit range and mobility. Second is the requirement for gearing systems: EM motors operate best at high rotational speeds; these must be reduced significantly through the use of gearing systems that can significantly increase mass and reduce energy density. Third is the ability to recover energy: tendon and flesh, as well as muscle itself, are capable of absorbing and storing a large percentage of the impact energy that can be translated back to motion. Additionally, muscles possess other salient properties that allow them to operate as motors, brakes, springs, and struts, permitting better stability control and impact energy absorption [8]. EM motors also generate more noise and heat than natural muscle, which is not welcome for certain applications, and cannot be effectively operated in large magnetic fields.

Pneumatic systems operate linearly like natural muscle; pneumatic artificial muscles (McKibben artificial muscles) in particular are intrinsically compliant and can thus provide the "give" that natural muscle attains. Unfortunately, these systems require air compressors that are neither light nor small, and their response speed is limited by the ability to pump air into and out of the actuators.

Several "smart materials" have been proposed as artificial muscles. These include shape memory alloys (SMA), magnetostrictive alloys (MSA), and piezoelectrics [2, 9]. SMAs are capable of producing relatively large linear displacements and can be actuated relatively quickly using resistive heating. What limits their applicability to artificial muscle applications is the time it takes to cool the alloy and return to the rest position. In order to obtain good operating frequencies, the SMA must be actively cooled, increasing the bulk, complexity and cost of the system. Magnetostrictive alloys and piezoelectric ceramics both suffer from small strains and high stiffness. These materials are thus not particularly suited to artificial muscle applications.

Polymers present an interesting alternative to conventional technologies. They possess inherent compliance, are lightweight, and are generally low cost. Electroactive polymers (EAPs) are an emerging type of actuator technology wherein a lightweight polymer responds to an electric field by generating mechanical motion [1, 10, 11]. Their ability to mimic the properties of natural muscle has garnered them the moniker "artificial muscle," though the term electroactive polymer artificial muscle (EPAM) is more appropriate and descriptive.

The concept of electroactive polymers can be dated back to 1880 in a paper by Roentgen [12]. In his experiments, he observed that a film of natural rubber could be made to change in shape by applying a large electric field across it; this was the first observation of actuation of a dielectric elastomer material.

Today the number of electroactive polymers has grown substantially. There currently exists a wide variety of such materials, ranging from rigid carbon-nanotubes to soft dielectric elastomers. A number of reviews and overviews have been prepared on these and other materials for use as artificial muscles and other applications [1, 2, 7, 10, 11, 13–28]. The next section will provide a survey of the most common electrically activated EAP technologies and provide some pertinent performance values. The remainder of the paper will focus specifically on dielectric elastomers. Several actuation properties for these materials are summarized in Table 1.1 along with other actuation technologies including mammalian muscle. It is important to note that data was recorded for different materials under different conditions so the information provided in the table should only be used as a qualitative comparison tool.

1.2 Survey of EAPEAP Technologies

EAPs can be broadly divided into two categories based on their method of actuation: ionic and field-activated. Further subdivision based on their actuation mechanism and the type of material involved is also possible. Ionic polymer-metal composites, ionic gels, carbon nanotubes, and conductive polymers fall under the ionic classification. Ferroelectric polymers, polymer electrets, electrostrictive polymers, and dielectric elastomers fall under the electronic classification.

1.2.1 Ionic Electroactive Polymers (EAPs)

1.2.1.1 Ionic Polymer-Metal Composites

Ionic polymer-metal composites (IPMCs) consist of a solvent swollen ion-exchange polymer membrane laminated between two thin flexible metal (typically percolated Pt nanoparticles or Au) or carbon-based electrodes [1, 29, 30]. Application of a bias voltage to the device causes the migration of mobile ions within the film to the oppositely charged electrode, causing one side of the membrane to swell and the other to contract resulting in a bending motion [31]. Over time the actuator will relax slightly due to the built-up pressure gradient. A schematic representation of the actuation mechanism is shown in Fig. 1.1. Typical membrane materials include Nafion and Flemion [20] with anionic side groups or polystyrene ionomers with anionic-substituted phenyl rings [1, 2, 29]. Wang et al. have recently shown that IPMCs based on a sulfonated poly(styrene-b-ethylene-co-butylene-b-styrene) ionic membrane are capable of high speed bending actuation under constant voltages and give excellent harmonic responses under

Table 1.1 Comparison of actuator materials

Type (Specific)	Maximum strain (%)	Maximum pressure (MPa)	Specific elastic energy density (J g^{-1})	Elastic energy density (J cm^{-3})	Coupling efficiency k^2 (%)	Maximum efficiency (%)	Specific density	Relative speed (full cycle)	Ref.
Dielectric elastomer (acrylic with prestrain)	380	7.2	3.4	3.4	85	60–80	1	Medium	[2, 3, 166]
Dielectric elatomer (silicone with prestrain)	63	3	0.75	0.75	63	90	1	Fast	[166]
Dielectic elastomer (silicone—nominal prestrain)	32	1.36	0.22	0.2	54	90	1	Fast	[3]
Bistable electroactive polymer (PTBA)	335	3.2	1.2	1.2	–	–	1	Medium	[139, 140]
Electrostrictive Polymer (P(VDF-TrFE))	4.3	43	0.49	0.92	–	~80 (est.)	1.8	Fast	[3]
Electrostatic devices (integrated force array)	50	0.03	0.0015	0.0025	50 (est.)	>90	1	Fast	[3, 4, 166]
Electromagnetic (voice coil)	50	0.1	0.003	0.025	–	>90	8	Fast	[3, 4]
Piezoelectric ceramic (PZT)	0.2	110	0.013	0.1	52	>90	7.7	Fast	[3]
Piezoelectric single crystal (PZT-PT)	1.7	131	0.13	1	81	>90	7.7	Fast	[3]
Piezoelectric polymer (PVDF)	0.1	4.8	0.0013	0.0024	7	–	1.8	Fast	[3]
Shape memory alloy (TiNi)	>5	>200	>15	>100	5	<10	6.5	Slow	[3]

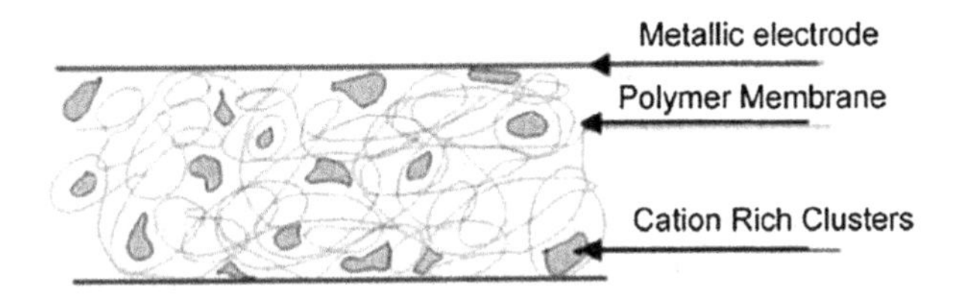

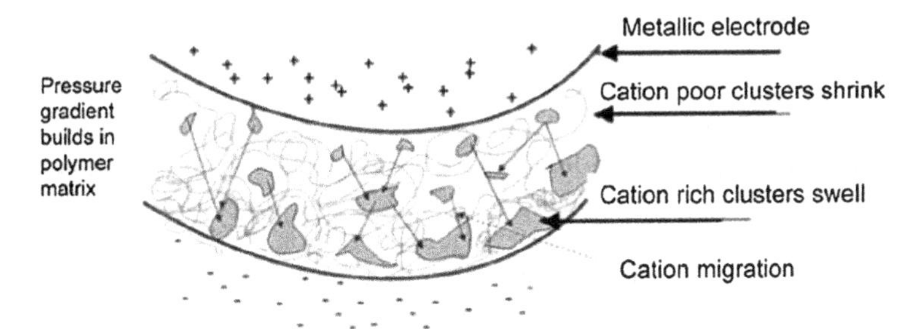

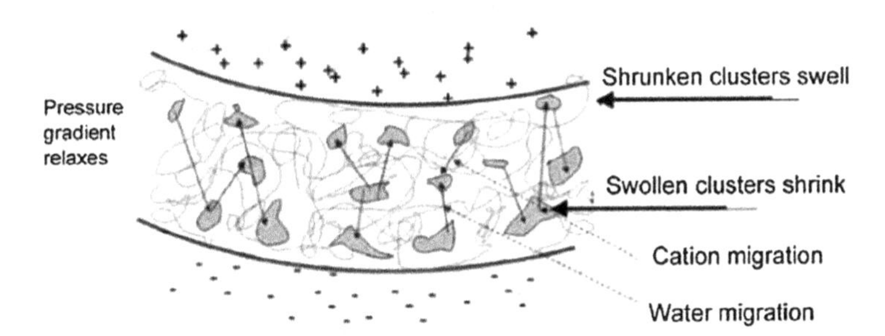

Fig. 1.1 Schematic representation of the actuation mechanism for an IPMC actuator. Application of a bias voltage causes mobile ions to migrate to one of the electrodes. The concomitant migration of solvent causes the ion rich region to swell, generating a bending motion. Over time the actuator relaxes due to the built-up pressure gradient [2]. IEEE 2004, reprinted with permission

sinusoidal excitation [32]. Since the membrane materials contain anionic species, they will be negatively charged, so cationic species are added to the solvent in the membrane to balance the charge. The ionic segments of the chains form hydrophilic clusters whereas the surrounding areas are hydrophobic; as such, the mobile ions accumulate near the ionic segments. Channels through the hydrophobic regions allow for ion and solvent migration [33]. Driving voltages are typically on the order of a few volts or less and actuation strains and stresses of >3% [30, 34] and 30 MPa [29, 33] have been reported. Several studies have demonstrated that IPMCs are well suited for use as soft actuators for bending and sensing [35–37]. Potential applications include mechanical grippers, metering valves, micropumps, and sensors [1, 30, 38]. Eamax, Japan has developed a commercially available fish robot that uses IPMC actuators [39]. Due to the low strains and the nature of their actuation, their applications for artificial muscle may be limited.

1.2.1.2 Ionic Gels

Ionic polymer gels consist of a crosslinked polymer, typically a polyacrylic gel acid, in an electrolyte solution. These materials are a class of hydrogels, a type of network polymer that swells in water. Hydrogels have been of interest for use as actuators for some time [40–43]. A hydrogel placed into an aqueous solution can change in shape and volume by a change in the polymer-liquid interaction, and hence the degree of swelling. This change can be brought about by a number of stimuli; the mostly commonly used is a change in pH. Polyacrylic gel acids will ionize in response to an increase in the pH, causing them to swell [44]. The change in pH can be induced by chemical means; however, this approach is impractical as it relies on a fluid-pumping system. Ionic gels also respond to electrical fields [45]. Application of an electric field to the gel causes the migration of hydrogen ions out of or into the gel resulting in a change in pH. The change in pH results in a reversible shift between swollen and contracted states. Actuators tend to bend in response to a DC field, which is caused by the difference between ion diffusion rates in the gel and in the electrolyte solution [46]. While bending may be useful for some applications, it is not particularly useful for artificial muscle applications. Calvert and Liu have reported on the swelling of layered gels, consisting of crosslinked polyacrylamide layers stacked on polyacrylic acid layers [47, 48]. When the pH is decreased to create a basic environment, the polyacrylic acid layers swell strongly; however, the polyacrylamide layers do not. The result is that the stacks expand with only marginal bending. Since the actuation depends on the diffusion of ionic species and the presence of a liquid electrolyte, actuation rates tend to be slow and encapsulation is an issue. These materials are still in the exploratory stages as artificial muscles. Recent work by Tondu et al. has probed the possibility of combining ionic gels with McKibben artificial muscles [49]. The intention is to replace the conventional pneumatic system with a chemical actuation mode with the goal of improving the response speeds and reducing system complexity.

1.2.1.3 Carbon Nanotubes

Since their discovery by Iijima [50], carbon nanotubes (CNTs) have garnered a great deal of interest thanks to their intrinsic mechanical and electrical properties and the ability to functionalize them and incorporate them in composite materials.

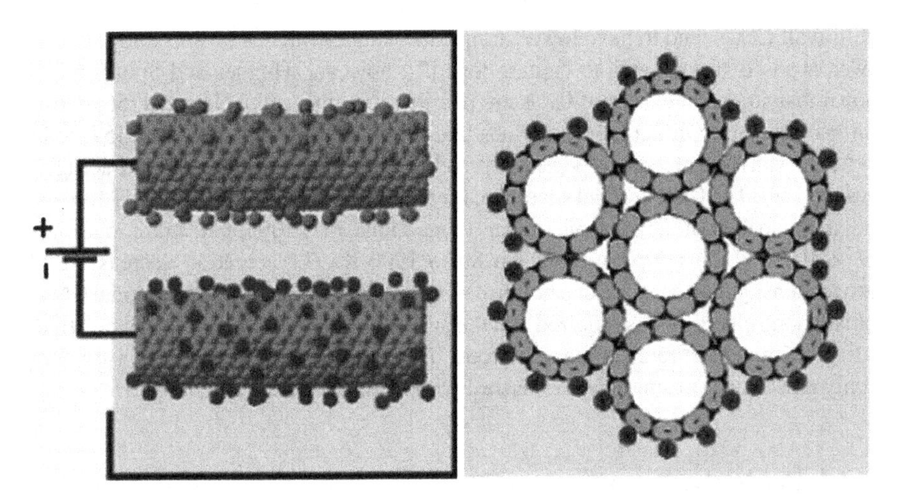

Fig. 1.2 Schematic representation of the actuation mechanism for a CNT actuator. When a bias is applied to CNTs that are submerged in an electrolyte, ions will migrate to the surface of the CNTs, which is offset by the rearrangement in their electronic structure. This phenomenon, coupled with Coulombic effects, results in actuation [7]. Materials Today 2007, reprinted with permission

Individual CNTs possess a high tensile modulus near that of diamond (640 GPa) and their tensile strength is thought to be 20–40 GPa, an order of magnitude larger than any other continuous fiber [51]. The mechanical properties of CNT bundles typically used in actuator tests tend to be much lower since they are held together by relatively weak van der Waals forces [52]. CNTs suspended in an electrolyte are capable of expanding in length due to double-layer charge injection [2]. When a bias is applied between the CNTs and a counter electrode, ions migrate to the surface of the CNTs. The resulting charge buildup must be offset by a rearrangement in the electronic charge within the tubes. The resulting actuation is due to these effects and to Coulombic forces [2, 24]. Figure 1.2 shows a schematic representation of the actuation mechanism. At low charge injections the quantum mechanical charge redistribution effects can predominate while at moderate to high levels the Coulombic effects dominate [52]. This mechanism does not require ion intercalation so lifetime and actuation rate are higher than for most ionic EAPs. CNTs possess low actuation voltages ($\sim$1 V), high operating loads (26 MPa), high effective power to mass ratios of 270 W Kg^{-1}, and a response speed in the millisecond range [53]. They should also be capable of work densities per cycle that are higher than any other current actuation material due to their high modulus. Strains are typically $<2\%$ [2] since CNTs are stiff and as such they would require strain amplifiers to be used as artificial muscles. Creep is also an issue, and can negatively affect the measured work densities [54]. CNTs also suffer from very poor electromechanical coupling. This can be partly attributed to the disparity in modulus between individual nanotubes and nanotube bundles [48]. Related work on polymer nanofibers has shown that the elastic modulus increases exponentially from bulk materials to nanofibers with diameters in the range of tens of nanometers [55].

Multi-wall CNTs tend to have lower strains than single-wall CNTs since they have a lower exposed surface area to capture ions [2]; however, Hughes and Spinks have shown that strains in excess of 0.2% are possible in multi-wall CNT mats [56]. Cost and manufacturing difficulties are issues that are currently being addressed [57, 58].

Aliev et al. have recently reported on novel giant-stroke, super-elastic CNT aerogel muscles [59]. The CNT aerogel sheets are fabricated from highly ordered CNT forests and are capable of anisotropic linear elongations of 220% and strain rates of $3.7 \times 10^4\% \ s^{-1}$ at temperatures from 80 to 1900 K. The actuation decreases the aerogel density and the strain can be permanently frozen in. Unlike conventional CNT actuators, no electrolyte is required and the actuation results from applying a positive voltage with respect to a counter electrode. The actuators have gas-like density and highly anisotropic mechanical properties, with Poisson's ratios reaching 15.

1.2.1.4 Conductive Polymers

Actuators based on conductive polymers (CPs) were first proposed by Baughman et al. in 1990 [60–62]. Much of the exploratory work was subsequently reported by Pei and Inganäs, and Smela, Inganäs and Otero et al. [63–66]. CPs actuate due to the uptake of counter-ions during electrochemical redox cycling, with the majority of the expansion occurring perpendicular to the polymer chain direction, indicating that ions and solvent are incorporated between the polymer chains [61, 67–69]. Changes in oxidation state result in a charge flux along the polymer chain and counter ions in the electrolyte migrate to balance the charge [1, 2, 7, 20, 69, 70]. For polymers doped with moderately sized counter anions, the oxidized state causes the polymer chains to expand due to the presence of the counter anions; in the reduced state the ions migrate away and the polymer chains relax. In polymers doped with bulky counter anions, the reduction involves uptake of cations into the polymer. The reduced polymer is in the expanded state [69–71]. Since both states are stable, these actuators are bistable. The most commonly used CPs for actuation purposes are polypyrrole and polyaniline [64, 72]. Figure 1.3 shows a schematic representation of a polypyrrole chain in its oxidized and reduced state. Actuation voltages are typically low (1–2 V) [63] and the materials are biocompatible. Strains are typically a few percent, but can range from 1 to ~40% [2, 7, 73–77]. Force densities up to 100 MPa are achievable [78] and values up to 450 MPa may be possible [63]. Conductive polymers suffer from a very low operating efficiency of ~1% and electromechanical coupling under 1% [2]. In addition, actuation speeds are limited since the mechanism relies on the migration of ions [20] and due to internal resistance between the electrolyte and polymer [4]. Due to the requirement of an electrolyte, most CP actuators must be encapsulated [79], though work has been performed on developing solid electrolytes that would remove this restriction [80]. Application of these materials for artificial muscles may be limited due to their poor coupling efficiencies and relatively slow actuation speeds (which would be exacerbated in large devices), though other potential biomimetic and human-interface applications exist that include blood vessel reconnection, dynamic Braille, valves, and catheters [81–83].

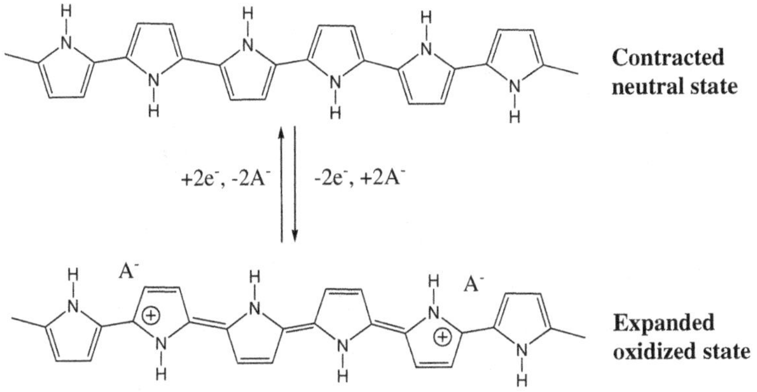

Fig. 1.3 Schematic representation of a polypyrrole chain in its oxidized and reduced states. Actuation results from the intercalation and deintercalation of ions between the chains

1.2.2 Field Activated EAPS

1.2.2.1 Ferroelectric Polymers

Ferroelectric polymers have a non-centro-symmetric structure that exhibits permanent electric polarization. These materials possess dipoles that can be aligned in an electric field and maintain their polarization. The induced polarization can be removed by applying a reverse electric field or by heating above the material's Curie temperature. They exhibit non-linear polarization curves demonstrating pronounced hysteresis. The polymers exhibiting these properties are limited mainly to poly(vinylidene difluoride) (PVDF), some PVDF copolymers, certain odd-numbered polyamides such as Nylon 7 and Nylon 11 [84], and blends thereof [85, 86]. In order to display ferroelectric behavior, polymers not only require polar side groups, they must also maintain molecular configurations in which the polarity does not cancel out. Thus, polymers such as poly(vinyl chloride) that has a polar carbon-chloride bond will not display ferroelectric behavior as it must arrange itself in a helical conformation due to steric effects from the relatively large van der Waals radius of the chloride atoms. Additionally, the polymer chains must be able to crystallize in a manner in which the polarization does not cancel. As an example, PVDF has four crystal structures [87, 88], a non-polar alpha phase, its polar analog (the delta phase), a highly polar beta phase, and a polar gamma phase. The morphology of the crystals must also be considered as they can have pronounced effects on the Curie temperature, remnant polarization, and other properties.

For commonly used poly(vinylidene fluoride trifluoroethylene), abbreviated as P(VDF-TrFE), copolymers with a VDF content ranging from 50 to 85 mol%, the ferroelectric beta phase is stable at room temperature, and a transformation to the paraelectric alpha phase occurs above the Curie point but below the melting point [89]. Above the Curie temperature, a transition between the paraelectric and

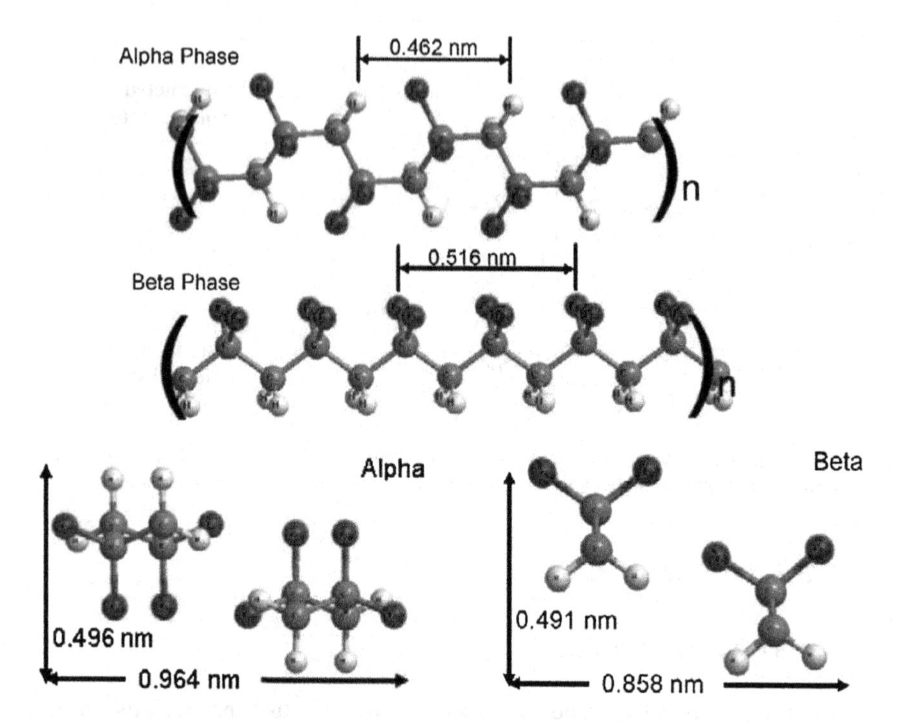

Fig. 1.4 The alpha and beta phases of the ferroelectric polymer PVDF. The beta phase is stable at room temperature but can be reversibly changed to the alpha phase by heating above the Curie temperature. Above the Curie temperature, an electric field can be used to induce a change between the alpha and beta phases. For ferroelectric relaxor polymers, the Curie temperature is below room temperature so the alpha phase is stable. A change to the beta phase can be induced by an electric field [7]. Materials Today 2007, reprinted with permission

ferroelectric phases can be brought about by applying an electric field [90, 91]. The change in phases results in an extremely large change in lattice constant, resulting in large bulk strains [92]. The Curie point decreases with decreasing crystallite size and can also be influenced by mechanical stress [18]. The induced change between the ferroelectric beta phase and paraelectric alpha phase is represented schematically in Fig. 1.4.

These materials have shown piezoelectric responses after appropriate poling [18]. Their piezoelectric actuation properties are typically worse than ceramic piezoelectric crystals; however, they have the advantages of being lightweight, flexible, easily formed, and not brittle. Additionally, while ceramics are limited to strains on the order of 0.1%, ferroelectric polymers are capable of strains of 10% [91] and very high electromechanical coupling efficiencies [93].

Recent advances in PVDF-based materials have led to the elimination of the hysteretic behavior characteristic of ferroelectrics. For this reason, these PVDF-based materials are classified as relaxor ferroelectric polymers; they will be discussed under the "Electrostrictive Polymers" heading.

1.2.2.2 Polymer Electrets

Early work by Eguchi on wax electrets [94] paved the way to the development of commercially viable low cost polymer electrets, with applications including microphones, sensors, transducers, and filters [18]. Electrets are insulating materials that display piezoelectric effects due to a non-uniform space charge distribution [95, 96]. Modern polymer electrets consist of a highly porous polymer with a polarization gas in the pores. The porous films are subjected to corona charging with voltages ranging from 5 to 10 kV. It is generally accepted that electrical discharging within the pores results in the build-up of charges at the polymer-gas interface. Positive and negative charges will lie on opposite sides of the pores according to the direction of the applied field, forming macroscopic dipoles [18]. Metal electrodes are applied to both sides of the film to act as contacts.

Polymer electrets can be operated as sensors or actuators. Their operation is very similar to that of a piezoelectric material and their direct piezoelectric transducer coefficient (d_{33}) is higher than that of solid PVDF ferroelectric polymers [97]. If a compressive force is applied to the film, the pores will deform preferentially with respect to the polymer material. Unlike charges within the polymer will be pushed closer together and the potential measured at the contacts will change accordingly. Similarly, the application of a voltage across the electrodes will yield a change in thickness in the material.

In order to meet increasing performance demands for electret applications, polymer electret blends are being explored. Lovera et al. have recently reported on tailored polymer electrets based on poly(2,6-dimethyl-1,4-phenylene ether) (PPE) and its blends with polystyrene (PS) [98]. They obtained good electret performance with neat PPE and showed that it could be improved by blending with PS.

1.2.2.3 Electrostrictive Polymers

Electrostrictive polymers have a spontaneous electric polarization. Electrostriction results from the change in dipole density of the material. These polymers contain molecular or nanocrystalline polarizations that align with an applied electric field. PVDF copolymers with nano-sized crystalline domains, electrostrictive graft copolymers, and liquid crystal elastomers fall under this category.

Relaxor Ferroelectric Polymers

Relaxor ferroelectric polymers are intimately related to the ferroelectric polymers described above. All known relaxor ferroelectric polymers are based on the P(VDF-TrFE) copolymer. As the name suggests, these polymers behave as relaxor ferroelectrics, which is distinguished by a broad peak in dielectric constant and a strong frequency dispersion [99, 100]. There are two major limitations of the P(VDF)-based ferroelectric actuators. First, the electrically induced paraelectric-ferroelectric transition that allows for actuation only occurs at temperatures above the Curie temperature, which is usually above room temperature. Second, the existence of strong hysteresis can make actuation more energy intensive and difficult to control. The Curie temperature of P(VDF-TrFE) copolymers can be lowered by introducing defects into the material, thereby reducing the size of the crystallites in the solid PVDF copolymer [18]. Having smaller crystallites also lowers the energy barrier required for the transition between paraelectric and ferroelectric states, which results in lower hysteresis [101]. For efficient room temperature operation, the Curie temperature must be reduced to around room temperature and the hysteresis must be suppressed. Zhang and others have shown this can be achieved by irradiating P(VDF-TrFE) with high-energy electrons or protons [101–104] or adding bulky side groups to the copolymer [102, 105–107], thus introducing polarization defects that destabilize the ferroelectric phase. Bulky side groups can be introduced through the formation of copolymers containing PVDF, TrFE and either chloride containing monomers such as cholorofluoroethylene (CFE) [102, 105] and chlorotrifluoroethylene (CTFE) [106, 107] or hexafluoropropylene (HFP) [108–110]. The actuation mechanism is essentially the same as for ferroelectric polymers; a transition between paraelectric and ferroelectric phases is induced by the application of a high electric field as represented in Fig. 1.4.

Recent work by Bao et al. has shown that P(VDF-TrFE) synthesized via reductive dechlorination from P(VDF-CTFE) exhibits ferroelectric relaxor behavior at high temperature ($\sim$100°C) with a melting point near 200°C [111]. This result is important as it provides another avenue to study the relaxor phenomena which are still not completely understood. The high melting point coupled with the high dielectric response of these materials at high temperature makes them attractive for use in high-temperature capacitors.

These materials have shown thickness strains on the order of 5% with fast response times [18]. Representative results are shown in Fig. 1.5. As seen in the figure, the required fields are quite high as in most electronic EAPs. Strains tend to show a peak and decrease for stresses above and below the peak value. Reported values for irradiated P(VDF-TrFE) show peak strains at 20 MPa, dropping to 50% of the maximum value above 40 MPa and below 5 MPa [112]. Elastic moduli in the range of 0.3–1.2 GPa have been reported with energy densities around 1 MJ m^{-3} [7]. In order to compete as artificial muscles, strain values will have to be improved.

Electrostrictive Graft-Copolymers

Electrostriction has also been obtained from graft copolymers wherein polar crystallites are grafted to flexible polymer backbones. The polar side groups aggregate to form crystalline regions which serve as the polarizable moieties required for actuation and as physical crosslinking sites for the flexible polymer as shown schematically in Fig. 1.6 [113]. When an electric field is applied to the copolymers, the polar crystallites reorient themselves which results in bulk deformation of the material. Strains and energy densities can be as high as 4% and 247 J kg^{-1} [114]. Similar results have been reported for a graft copolymer consisting of chlorofluoroethylene

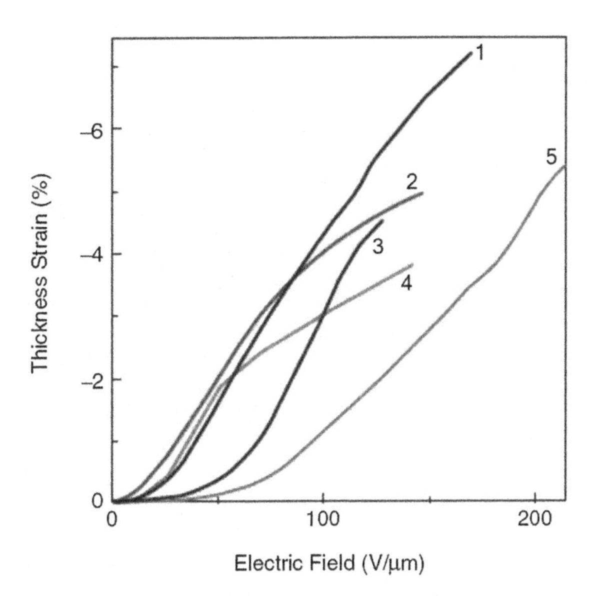

Fig. 1.5 Actuation strain as a function of electric field for irradiated P(VDF-TrFE) copolymer (2), P(VDF-TrFE-CTFE) (4), P(VDF-CTFE) (5), and P(VDF-TrFE-CFE) (1 and 3, the two curves have different compositions) [18]. Materials Research Society 2008, reprinted with the permission of Cambridge University Press

and trifluoroethylene backbone with P(VDF-TrFE) side chains [1]. While the required electric fields are lower than for relaxor ferroelectrics, the strains are lower and the actuation rates are slower due to the size of the polar crystallites. Unimorph and bimorph bending actuators have been demonstrated [114].

Liquid Crystal Elastomers

As the name suggests, liquid crystal elastomers (LCEs) combine the orientational ordering properties of liquid crystals with the elastic properties of elastomer networks [115]. As early as 1975, de Gennes predicted that the reorientation of mesogens in liquid crystals during a phase transition could result in bulk stresses and strains [116]. LCEs were first proposed for use as artificial muscles by de Gennes in the late 1990s [117]. LCEs consist of mesogens attached to one another via a networked elastic polymer. The polymer network permits sufficient motion to allow for the rotation of the mesogens while maintaining a solid shape and preventing free flow. LCEs can be divided into two categories depending on the phases present: nematic and smectic. The actuation mechanism for these two systems is different. Here we will briefly describe the nematic system since it is the focus of the majority of the current research in the field; the interested reader can consult references [118–120] and [121] for a detailed look at the nematic and smectic systems respectively. In nematic polymer systems where the mesogens are incorporated directly into the backbone, the chains will elongate in their nematic phase when all of the mesogens are aligned [122–124] and will relax when the polymer is in the isotropic state and the chains are allowed to coil up into their entropically favored positions [125]. Similar effects are observed when the

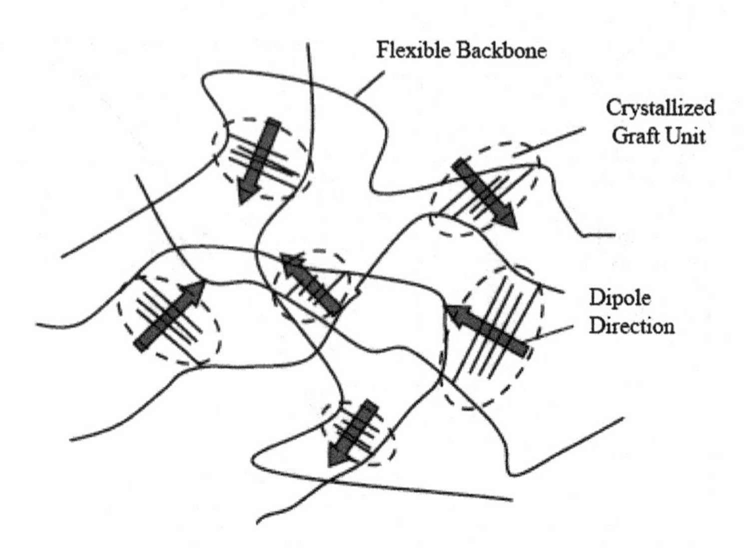

Fig. 1.6 Electrostrictive graft copolymer consisting of P(VDF-TrFE) main chains and PVDF grafts. Polar PVDF units (*pink*) aggregate and form crystallites that serve as polarizable moieties for actuation and as physical crosslinks between elastomer chains [113]. J. Su 2003, reprinted with permission

mesogens are attached as grafts to the elastomeric backbone. The change in orientation can be effected via thermal and electrical stimuli [116, 118, 121, 126–132]. Thermally activated LCEs display length changes as high as 400%, but their response speed is limited due to the requirement for heat diffusion [117, 133–134]. Several strategies have been investigated for improving the thermal diffusivity [129, 135] of LCEs, but thermal relaxation is still an issue. The thermal response can be generated from a number of stimuli, including optical, electrical, and direct magnetic through the incorporation of appropriate additives [136]. Electrically activated LCEs have intrinsically polarized mesogens that can realign in the presence of an electric field to generate bulk stress and strain as shown schematically in Fig. 1.7 [116]. Electrically activated LCEs have much faster response speeds (10 ms) [121] than the thermally activated variety and the required fields (1.5–25 MV m^{-1}) are lower than for most other field activated EAP technologies; however, the actuation strains are relatively small (<10%). Recent results have shown 4% strain at 133 Hz with a field strength of 1.5 MV m^{-1} [121]. The combination of small to moderate strains with low moduli means that the work densities of these materials will be relatively low. The use of stiffer polymers yielded strains of 2% at 25 MV m^{-1} and a work density of 0.02 MJ m^{-3} [131]. Additional improvements in strain and work density will be required if LCEs are to compete as artificial muscles. Recent results have shown that the application of a high electric field across a smectic LCE results in a large electroclinic effect with reasonable rates at relatively low voltages that can be used for actuation [137, 138].

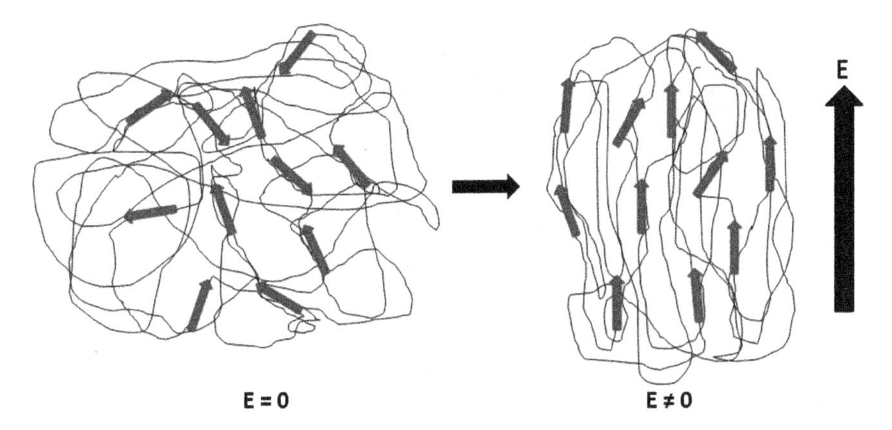

Fig. 1.7 Actuation mechanism of a liquid crystal elastomer. The application of an electric field results in the realignment of intrinsically polarized liquid crystal mesogens. The mesogens are either grafted to elastomer chains or incorporated within them. The elastomer chains prevent the free flow of the mesogens and couple their motion to bulk stresses and strain

1.2.2.4 Bistable Electroactive Polymers

Bistable electroactive polymers (BSEPs) are a new class of electroactive polymer developed in 2009 by Yu et al. [139]. BSEPs combine the properties of dielectric elastomers (described in the following section) and shape memory polymers (SMPs). As with SMPs, BSEPs behave as rigid plastics below their glass transition temperature (T_g) and soft elastomers above their T_g. If a thin film of the material is sandwiched between compliant electrodes, it can behave as a capacitor in the rigid state and as a compliant variable capacitor in the elastomeric state. If a voltage is applied across the BSEP film in the softened state, the resulting electrostatic forces will act to compress the film in thickness and expand it in area. In the rigid state, the modulus of the film is too large to be electrostatically deformed. If the film is actuated in the softened state, then cooled to below its T_g, it will retain the actuated shape indefinitely without the requirement for an applied voltage. In addition, because of the transition to a rigid glassy state, the material is able to support relatively large loads. When reheated, the film returns to its rest position. Because the strain is electrically controlled, BSEPs can be reversibly actuated and locked into any position between the rest state and their maximum deformation state. Figure 1.8 shows an example of a BSEP being actuated in a diaphragm configuration. The film is heated to 70°C and actuated to a high strain state; it is then cooled to room temperature to lock in the strain and the driving voltage is removed. Upon reheating to 70°C, the film returns to its original state.

Poly(tert-butyl-acrylate) (PTBA) has been extensively studied as a BSEP material [139, 140]. It exhibits a relatively sharp glass transition at 50°C. At room temperature PTBA has a storage modulus of 1.5 GPa and a loss factor, tan δ, of 0.03. At 70°C, the bulk modulus is reduced to 0.42 MPa with a loss factor of 0.8.

Fig. 1.8 Diaphragm actuator with a BSEP film and conductive carbon grease coated on both surfaces as the compliant electrodes. (**a**) Initial device at room temperature or 70°C; (**b**) Applying 1.8 kV at 70°C followed by cooling to room temperature and removal of the actuation voltage; (**c**) raising temperature to 70°C without any external voltage being applied [139]. Appl Phys Lett 2009, reprinted with permission

PTBA also exhibits excellent strain fixity (ability to retain its actuated shape upon cooling) and strain recovery. In its softened state PTBA also possesses excellent actuation properties with a breakdown field strength in excess of 250 MV/m, a maximum strain of 335% in area, a maximum actuation stress of 3.2 MPa and an energy density of 1.2 J cm^{-3}, values that rival even the best of the conventional dielectric elastomer materials. The BSEP is the first active material that possesses bistable actuation with high strain and specific power density.

The ability to lock in strain is very important for applications wherein the device must hold its actuated state for an extended period of time. Conventional dielectric materials consume energy when actuated due to current leakage through the film, and can succumb to premature breakdown when held at high strain for an extended period of time. By locking in the actuated shape, BSEPs can hold their actuated shape without draining power and can maintain that shape indefinitely without failure. This combination of properties places BSEP materials at the forefront in terms of electroactive polymer materials for artificial muscle applications.

1.2.2.5 Dielectric Elastomers (DEs)

Dielectric elastomer (DE) actuators are essentially compliant variable capacitors. They consist of a thin elastomeric film coated on both sides by compliant electrodes. When an electric field is applied across the electrodes, the electrostatic attraction between the opposite charges on opposing electrode and the repulsion of the like charges on each electrode generate stress on the film causing it to contract in thickness and expand in area (Fig. 1.9). Most elastomers used are essentially incompressible, so any decrease in thickness results in a concomitant increase in the planar area.

Typical operating voltages for DE films 10–100 μm in thickness range from 500 V to 10 kV. The area expansion can be readily measured if the films are subjected to tensile prestrain: the non-active areas in tension surrounding the active area pulls the expanded active area and keeps it flat (Fig. 1.10).

The driving currents are very low and the device is electrostatic in nature, so it will theoretically only consume power during an active area expansion (thickness

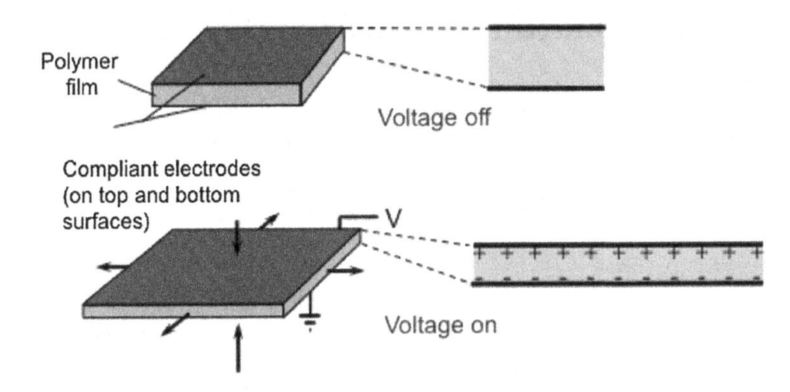

Fig. 1.9 Dielectric elastomer operating principle. When a bias voltage is applied across an elastomer film coated on both sides with compliant electrodes, Coulombic forces act to compress the film in the thickness direction and expand it in plane [1]. © SPIE Press 2004, reprinted with permission

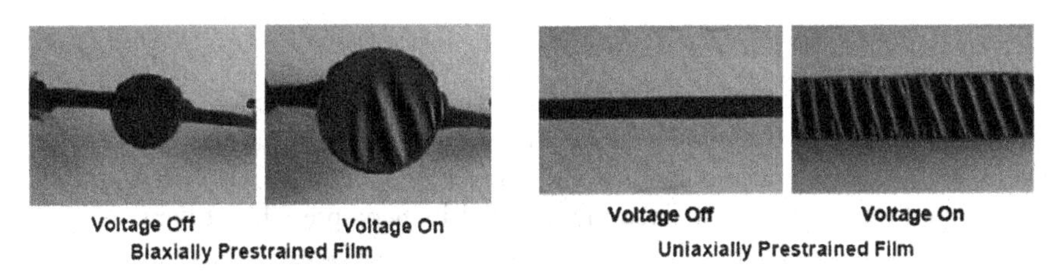

Fig. 1.10 Actuation of dielectric elastomer devices with biaxial and uniaxial prestrain. Uniaxial prestrain results in preferential in-plane strain in the direction perpendicular to the applied prestrain direction [1]. © SPIE Press 2004, reprinted with permission

reduction) and no power will be consumed to maintain the DE at a stable actuated state. Furthermore, some of the energy can be recovered after the actuation cycle is complete. In practice, however, there will be some leakage current through the dielectric, the amount of which will depend on the material and its thickness, and thus the DE will consume a small amount of power when maintained in a stable actuation state. Viscoelastic effects may also play a role in reducing efficiency.

Output stress varies quadratically with electric field; however, due to the inherent compliance of the materials, the force that can be coupled to a load will decrease with increasing strain for a particular electric field. Maximum force is available at zero strain and at maximum strain the elastomer will not generate any output forces. Note also that for a given strain, the output force, and thus the stiffness, can be modulated by varying the applied field. This is an important feature for artificial muscle applications as it allows the DE actuators to "brace" themselves as natural muscles do to maintain stability or prevent damage.

1.3 Modeling of Dielectric Elastomer Materials

The actuation of DEs can be approximated as the lateral electrostatic compression and planar expansion of an incompressible linearly elastic material where the electrical component is treated as a parallel plate capacitor [141]. The incompressibility constraint can be expressed as:

$$Az = P \tag{1.1}$$

where A is the area of the electrodes, z is the thickness of the elastomer film between electrodes, and P is a constant. The stored electrical energy on the DE is given by the stored energy on a parallel plate capacitor:

$$U = 0.5\frac{Q^2}{C} = 0.5\frac{Q^2 z}{\varepsilon_r \varepsilon_0 A} \tag{1.2}$$

where Q is the charge on the electrodes, C is the capacitance, ε_r is the relative permittivity and ε_0 is the permittivity of free space. The electrostatic pressure across the electrodes is then given by the Maxwell pressure:

$$p = \varepsilon_r \varepsilon_0 E^2 \tag{1.3}$$

This is exactly twice the pressure across a parallel plate capacitor. The factor of two is due to the incompressibility of the elastomer film. Charges on opposite electrodes will attract one another, resulting in a reduction in thickness as well as a concomitant increase in area since the material is incompressible. Likewise, like charges on each electrode will also repel each other, causing an increase in area and a concomitant reduction in thickness.

Using the linear-elasticity and free boundary approximations used in the early work in the field, which is only valid for small strains (<10%), the change in thickness is given by [141]:

$$s_z = -\frac{p}{Y} = -\frac{\varepsilon_r \varepsilon_0 E^2}{Y} = -\frac{\varepsilon_r \varepsilon_0 (V/z)^2}{Y} \tag{1.4}$$

where V is the applied voltage and Y is the elastic modulus.

Krokovsky et al. provided a good derivation of this linear small-strain case and extended these derivations to include the effects of electrostriction that may be important for certain materials [142]. Pelrine et al. showed that for the small strain case, the actuator energy density is given by [3, 143]:

$$e_a = -p s_z = \frac{(\varepsilon_r \varepsilon_0)^2 E^4}{Y} = \frac{(\varepsilon_r \varepsilon_0)^2 (V/z)^4}{Y} \tag{1.5}$$

This equation considers that both the expansion and contraction in an actuation cycle can exert work. Similarly, the elastic energy density is given by:

$$e_e = -\frac{1}{2} p s_z = \frac{1}{2}\frac{(\varepsilon_r \varepsilon_0)^2 E^4}{Y} = \frac{1}{2}\frac{(\varepsilon_r \varepsilon_0)^2 (V/z)^4}{Y} \tag{1.6}$$

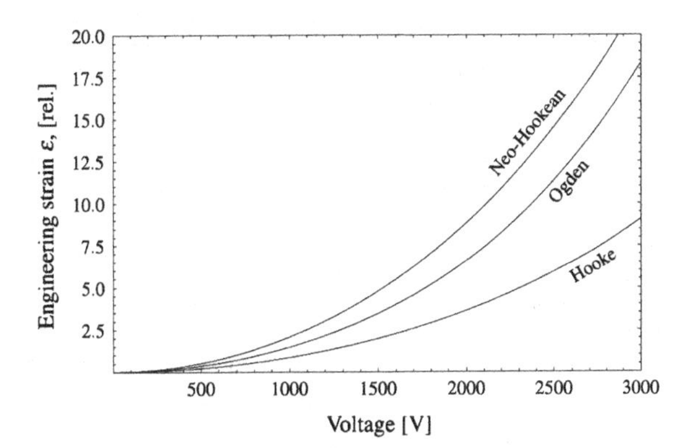

Fig. 1.11 DE voltage-strain curves as predicted by various models [147]. © G Kofod 2001, reprinted with permission

For larger strains, while maintaining the assumption that the material is linearly elastic, they showed that [141]:

$$s_z = \frac{2}{3} + \frac{1}{3}\left[f(s_{z0}) + \frac{1}{f(s_{z0})}\right] \tag{1.7}$$

where:

$$f(s_{z0}) = \left[2 + 27s_{z0} + \frac{\left(-4 + (2 + 27s_{z0})^2\right)^{1/2}}{2}\right]^{1/3} \tag{1.8}$$

and:

$$s_{z0} = -\varepsilon_r \varepsilon_0 \frac{V^2}{Y z_0^2} \tag{1.9}$$

In this case, the elastic energy density is [3, 143]:

$$e_e = Y[s_z - \ln(1 + s_z)] \tag{1.10}$$

The preceding equations provided a reasonable foundation for predicting DE behavior. Indeed the assumption that DEs behave electronically as variable parallel plate capacitors still holds; however, the assumptions of small strains and linear elasticity limit the accuracy of this simple model. More advanced non-linear models have since been developed employing hyperelasticity models such as the Ogden model [144–147], Yeoh model [147, 148], Mooney-Rivlin model [145–146, 149, 150] and others (Fig. 1.11) [147, 151, 152]. Models taking into account the time-dependent viscoelastic nature of the elastomer films [148, 150, 151], the leakage current through the film [151], as well as mechanical hysteresis [153] have also been developed.

Recently Zhao and Suo developed a thermodynamic model of electrostriction for elastomers capable of large deformation that helps elucidate the roles that different electrostrictive effects play at different strain levels [154]. Others have attempted to study dynamic effects for actuators subjected to time-varying voltage and pressure inputs [155, 156].

Models have been used to study the failure modes and instabilities present in DEs, including the thermodynamic electromechanical instability (pull-in) [157–163], wrinkling/buckling [158–159], mechanical rupture [159, 160], and dielectric breakdown [159, 160] for select actuator configurations. An improved understanding of these failure modes should allow for the design of actuators capable of operating in a safe regime, thus prolonging actuator lifetime.

The applicability of any of the models to a general case is suspect since the modeling parameters are strongly dependent on the testing conditions. Factors such as prestrain, mechanical loading, actuator configuration, humidity, and temperature can have a large effect on the parameters obtained. However, these studies have provided useful insight into the failure mechanisms present in DE actuators and provide tools for design engineers to develop new actuator configurations capable of large strains, high forces outputs, and long lifetimes.

1.4 Dielectric Elastomer Materials

In the late 1990s and early 2000s a large number of elastomer materials were tested, including silicones, polyurethanes, isoprene, and fluoroelastomers [141, 142, 164, 165]. Pertinent properties of these materials and several other dielectric elastomers described below are included in Table 1.2. Pelrine et al. identified three particularly promising groups of materials: silicones, polyurethanes, and acrylics. Their actuation characteristics were promising; however, it was not until strains in excess of 100% in area in both silicone and acrylic elastomer films were reported that significant interest was garnered in the scientific community [143, 166]. The key to developing such large strains in these materials was prestrain. Though the exact mechanism by which actuation strains are improved is not known, it has been shown that prestrain enhances the breakdown field in certain acrylic elastomers and can reduce viscoelastic effects.

Of these materials, a commercially available 3M VHB acrylic elastomer (VHB 4910 and VHB 4905) appears to be the most promising in terms of strain performance, with strains in excess of 380% reported for highly prestrained films [4]. The theoretical energy density of this elastomer is an impressive 3.4 MJ m^{-3} and coupling coefficients as well as efficiencies as high as 90% are possible.

Polyurethane films were pursued because of their larger force outputs and higher dielectric constant, allowing them to be actuated at lower electric fields. However, polyurethane films are limited in their ability to generate large strain.

Silicone elastomers have the advantage of lower viscoelasticity than acrylic films and can therefore be operated at higher frequencies with lower losses. Silicones show modest actuation strain when there is little to no prestrain and can be operated over a wide temperature range, making them more suitable to

Table 1.2 Comparison of dielectric elastomer material properties

Polymer (specific type)	Prestrain (x, y%)	Energy density (MJ • m^{-3})	Actuation pressure (MPa)	Thickness strain (−%)	Area strain (%)	Young's modulus (MPa)	Electic field (MV • m^{-1})	Dielectric constant[b]	Dielectric loss factor[b]	Mechanical loss factor	Coupling efficiency k^2 (%)	Efficiency (%)[d]	Ref.
Silicone (Nusil CF19-2186)	–	0.22[e]	1.36[f]	32	–	1	235	2.8	54	–	–	–	[3]
Silicone (Nusil CF19-2186)	(45, 45)	0.75[e]	3	39	64	1.0[g]	350	2.8	6.3	0.005	0.05	79	[143, 161]
Silicone (Nusil CF19-2186)	(15, 15)	0.091[e]	0.6	25	33	–	160	2.8	–	–	–	–	[143, 166]
Silicone (Nusil CF19-2186)	(100, 0)	0.2[e]	0.8	39	63 (linear)	–	181	2.8	–	–	–	–	[143, 166]
Silicone (Dow Corning HS3)	–	0.026[e]	0.13[f]	41	–	0.135	72	2.8	65	–	–	–	[3]
Silicone (Dow Corning HS3)	(68, 68)	0.098[e]	0.3	48	93	0.1[g]	110	2.8	79	0.005	0.05	82	[143, 166]
Silicone (Dow Corning HS3)	(14, 14)	0.034[e]	0.13	41	69	–	72	2.8	–	–	–	–	[143, 166]
Silicone (Dow Corning HS3)	(280, 0)	0.16[e]	0.4	54	117 (linear)	–	128	2.8	–	–	–	–	[143, 166]
Silcone (Dow Corning Sylgard 186)	–	0.082[e]	0.51[f]	32	–	0.7	144	2.8	54	–	–	–	[3]
Polyurethane (Deerfield PT6100S)	–	0.087[e]	1.6[f]	11	–	17	160	7	21	~0.5	~0.08 (@ 30 Hz)	–	[3, 205]
Polyurethane (Estane TPU588)	– –	0.0025	0.14	8	–	–	8 (at max. strain)	6	–	–	–	–	[209]
Polyurethane-Carbon Powder Composite (Estane TPU588)	–	0.0043	0.14	12	–	–	8 (at max. strain)	6	–	–	–	–	[209]
Fluorosilicone (Dow Corning 730)	–	0.0055[e]	0.39[f]	28	–	0.5	80	6.9	48	–	–	–	[3]
Fluoroelastomer (Lauren L143HC)	–	0.0046[e]	0.11[f]	8	–	2.5	32	12.7	15	–	–	–	[3]

(continued)

Table 1.2 (continued)

Polymer (specific type)	Prestrain (x, y%)	Energy density (MJ • m^{-3})	Actuation pressure (MPa)	Thickness strain (−%)	Area strain (%)	Young's modulus (MPa)	Electic field (MV • m^{-1})	Dielectric constant[b]	Dielectric loss factor[b]	Mechanical loss factor	Coupling efficiency k^2 (%)	Efficiency (%)[d]	Ref.
Isoprene Natural Rubber Latex	–	0.0059[e]	0.11[f]	11	–	0.85	67	2.7	21	–	–	–	[3]
Dr. Scholl's Gelactiv Tubing	(140, 0)	–	0.0037	1.8	–	–	28	–	–	–	–	–	[167]
Acrylic (3M VHB)	–	3.4[e]	7.2	79	380	–	–	–	–	–	–	60–80	[4]
Acrylic (3M VHB 4910)	(300, 300)	3.4[e]	7.2	61	158	3.0[g]	412	4.8	90	<0.005	0.18	80	[143, 166]
Acrylic (3M VHB 4910)	(15, 15)	0.022[e]	0.13	29	40	–	55	4.8	–	–	–	–	[143, 166]
Acrylic (3M VHB 4910)	(540, 75)	1.36[e]	2.4	68	215 (linear)	–	239	4.8	–	–	–	–	[143, 166]
Acrylic (3M VHB 4910)	Nominal	0.0057	–	7	7.5	2.3[b]	17	4.2	14.5	–	–	–	[203]
Acrylic (3M VHB 4905)	Nominal	0.0014	0	11	12.4	2.3[b]	34	4.2	20.9	–	–	–	[203]
SEBS161	(300, 300)	0.141–	–	62–22	180–	0.007–	32–133	1.8–2.2	92–53	–	–	–	[203]

[a] Breakdown field unless otherwise stated
[b] At 1 kHz
[c] At 80 Hz unless otherwise stated
[d] At 80 Hz
[e] Estimated via calculation
[f] Calculated Maxwell pressure
[g] Effective modulus
[h] Measured in compression

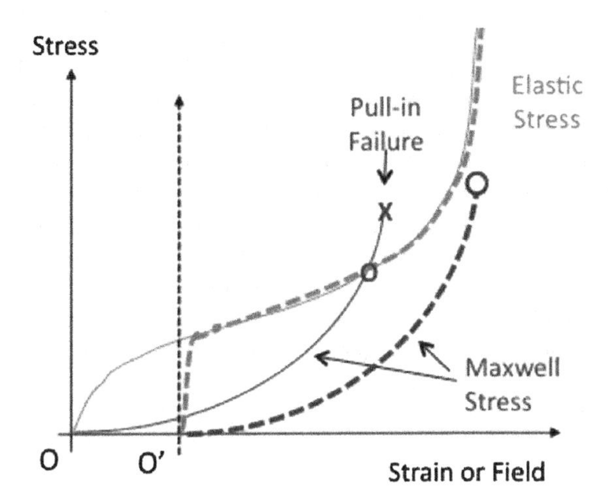

Fig. 1.12 Characteristic stress of a dielectric elastomer film as a function of mechanical strain or electric field (constant voltage condition). The charts with origin at O are for a non-prestrained film and at O′ for the prestrained film. The *cross* (X) indicates dielectric breakdown and the *bar* (—) indicates stable actuation strain. *Small* o and *large* O represent the apparent breakdown field and actual breakdown strength, respectively

applications where temperatures are expected to vary significantly. They are also capable of strains in excess of 100% when prestrained, but fall short of acrylics in this area. Silicones also possess a relatively low dielectric constant and thus require higher electric fields.

Carpi et al. have reported on the actuation characteristics of another commercially available elastomer (Dr Scholl's, Canada, Gelactiv tubing) [167]. The material was capable of thickness strains of 1.8% at 27 MV m^{-1} with an actuation stress of 3.7 kPa at 24 MV m^{-1}.

It follows that many of the most exciting dielectric elastomer materials were discovered via exploratory testing. New formulations of commercially available elastomers are continually being developed and may well be worth exploring. However it is expected that research focusing on developing materials specifically for dielectric elastomer purposes from focused and directed research will provide the best candidates for improved performance in the years to come.

1.4.1 Effects of Prestrain

It has been found that prestrain can significantly improve the actuation performance of dielectric elastomer devices [143, 165]. The observed improvements have been largely attributed to an increase in the breakdown strength [168–169], which has been explained via a thermodynamic stability criterion [170]. Prestrain has the additional benefits of improving the mechanical efficiency [171] and response speed of most dielectric elastomers while causing a marginal decrease in the dielectric constant [172]. Prestrain can also be used to achieve preferential actuation in a certain direction by applying high prestrain in the direction perpendicular to the desired actuation direction and low prestrain along the actuation direction [143].

The effects of prestrain on the actuation performance of dielectric elastomers have been widely studied from both a practical as well as theoretical perspective [143, 165, 168–182]; however, the mechanism by which prestrain increases dielectric strength is still not completely understood. It is generally believed that premature DE failure is exacerbated by localized defects introduced during manufacture, processing, or through fatigue. The localized defects act to reduce the local breakdown strength in that area which can lead to localized or permanent device failure. Localized pull-in effects, viscoelastic behavior, and high leakage current can also reduce the usable electric field. It is likely that prestrain prevents pull-in effects. As shown in Fig. 1.12 for a given material, a non-prestrained film will undergo a characteristic rapid increase in stress, then a plateau region, followed by a very steep rise in the stress until the film fails via the breaking of covalent bonds. The actuation stress/strain curve will follow the quadratic curve in the figure. For sufficiently high driving voltages, the reduction in film thickness and increase in electric field intensity form a positive feedback loop. The film will continue to be driven thinner and thinner until the local electric field exceeds the dielectric strength of the film, as indicated by the intersection of the quadratic curve and elastic stress curve. This induces the pull-in effect and possible eventual dielectric breakdown. If the film is prestrained, the origin of the actuation stress/strain curve will shift to O'. The resulting actuation stress/strain curve will be less likely to intersect the elastic stress/strain curve; therefore the actuation will be more stable. Prestrain is very effective for acrylic films and is also effective in silicone and other elastomers. The effect varies from material to material.

Prestrain has the additional benefit of improving the frequency response of many elastomer films. The drop off in actuation strain with frequency is less pronounced for acrylic films that have been prestrained. The increased tension in the film increases the modulus and reduces the viscoelastic nature of the films.

Kofod used advanced materials models in an attempt to elucidate the effects that prestrain have on the actuation performance of a simple cuboid DE actuator [183]. The results are purely phenomenological; however, they indicate that in the special case of a purely isotropic amorphous material, prestrain does not affect the electromechanical coupling directly. The enhancement in actuation strain due to prestrain occurs through the alteration of the geometrical dimensions of the actuator. Kofod also determined that the presence of an optimum load is related to the plateau region in the force–stretch curve and that prestrain is not able to affect the location of this region.

The unfortunate drawback to prestraining films is that a rigid frame, or other structure, must be used to maintain the tension in the film. The added mass of the supporting structure increases the total mass of the DE device, which can significantly reduce the effective work density and power to mass ratio of the actuator. The prestrained films may also relax or fatigue over time. This reduces the shelf life of the DE devices.

1.4.2 Improved Silicone Films

A considerable amount of research has focused on reducing the operating voltages of DE actuators in order to increase their commercial viability and remove the dangers associated with high voltage. There are two basic methods of reducing the operating voltage: reducing the thickness of elastomer films so that the required field for high-performance actuation occurs at lower voltages; and increasing the dielectric constant of the elastomer films to reduce the required electric field intensity. Reducing of film thickness has the benefit of maintaining the dielectric breakdown strength and dielectric loss of the film, but suffers from reduced output force and the increased importance of inhomogeneities that cause localized areas of high electric field and stress and result in premature breakdown. The output force can be scaled up through the use of multilayer actuators, but doing so increases fabrication complexity.

A number of approaches have been explored for increasing the dielectric constant of elastomers for DEs. The most common approach involves the addition of high dielectric constant filler materials to an elastomer host. Silicone is of particular interest for this type of approach as it possesses good actuation properties to begin with, is readily available in gel form, and has a low dielectric constant. Results thus far do not appear particularly promising: increases in dielectric constant have been met with concomitant increases in dielectric loss and reductions in dielectric breakdown strength [184–186]. It has also been shown that the elastic modulus is affected by the addition of filler [187].

Researchers have investigated the effects of adding various oxides including aluminum oxide, titanium oxide, and barium titanium oxide [186–189], as well as various other fillers including organically modified montmorillonite (OMMT) [190], lead magnesium niobate-lead titanate (PMN-PT) [185, 188], copper-phthalocyanine oligomer (CPO) [188, 191], and PEDOT/PSS/EG [192]. Improvements appear limited at best as any meaningful increases in the dielectric constant are achieved at loadings near the percolation threshold and are met with increases in the leakage current. Research is still ongoing and some researchers believe that with some additional refinement real improvements can be achieved.

Carpi et al. have recently developed silicone-based polymer blends that display enhanced electromechanical transduction properties [193]. Their technique involves blending, rather than loading, the silicone elastomer with a highly polarizable conjugated polymer. They have reported very promising results for silicones loaded with very low percentages (1–6 wt%) of poly(3-hexylthiophene) (PHT). The resulting blends yielded an increase in dielectric permittivity, with a relatively small increase in dielectric loss, and a reduction in tensile elastic modulus, which contribute synergistically to an improvement in electromechanical performance. The best performance was obtained for a blend with 1 wt% PHT, achieving a transverse strain of 7.6% at a field of only 8 MV m^{-1}.

Chen et al. have developed an electrothermally actuated CNT-silicone composite [194]. They have been able to produce a maximal strain of 4.4% with an applied field of only 1.5 kV m^{-1} by incorporating a CNT network into the silicone

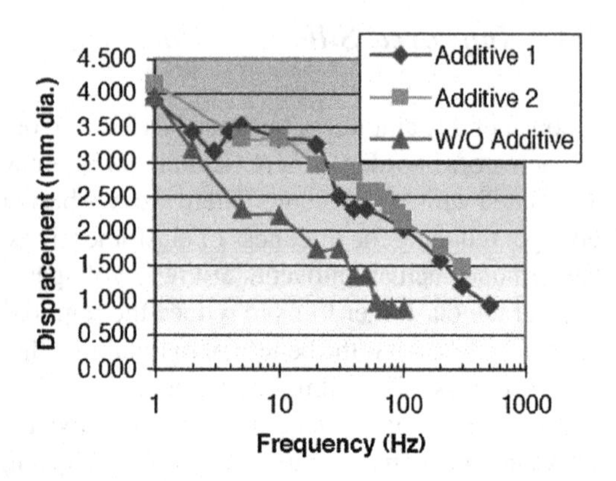

Fig. 1.13 Improved frequency response of VHB acrylic elastomers via the addition of low molecular weight additives (plasticizers) [195]. Proc SPIE 2004, reprinted with permission

elastomer. The mechanism of actuation differs strongly from that of conventional DEs. In these materials, the CNT loadings are such that conductivity through the specimen is possible. The flow of current through the CNT matrix causes joule heating, which raises the temperature of the silicone matrix, causing it to expand due to thermal effects.

1.4.3 Improved Acrylic Films

Conventional acrylic films, such as the VHB 4910 series of elastomers from 3M, possess excellent actuation strain, energy density, and coupling efficiency. However, in order to achieve these high performance values, the film must be prestrained. The addition of bulky support frames required to maintain the prestrain on the film significantly increases the mass of VHB acrylic based devices, reducing their effective energy densities to more pedestrian values. VHB acrylic films also suffer from viscoelastic effects, which limit their maximum response frequency to the 10–100 Hz range. The viscoelastic nature of these films also limits their overall efficiency and results in time dependent strain that can make their performance somewhat erratic.

Low molecular weight additives can increase the frequency response of VHB films [195]. Representative results are shown in Fig. 1.13. For the film without additives, the strain reduced to half its static value at 12 Hz; for the films with additives this frequency was pushed to over 100 Hz. These additives have the additional benefit of decreasing the glass transition temperature, thereby increasing the range over which the VHB films can be operated. When added in high concentrations, however, the additives reduce the mechanical stability of the films, making them easier to tear.

Interpenetrating polymer networks (IPNs) have been synthesized combining acrylic and silicone rubber materials [196]. These IPN films are synthesized by diffusing silicone chains into swollen acrylic rubber films in the presence of a co-solvent and then crosslinking the silicone chains. The resulting films display

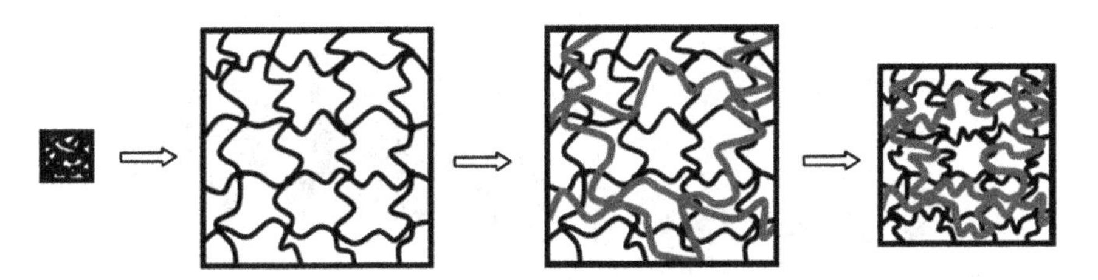

Fig. 1.14 Fabrication steps for IPN elastomer films. The film is first prestrained, then an multifunctional monomer additive is sprayed onto the host film and polymerized forming an interpenetrating polymer network. Upon releasing the film it retains most of the applied prestrain, with the additive network being in compression and the host film in tension [200]. Proc SPIE 2007, reprinted with permission

properties between those of acrylic and silicone films as expected. Such IPN films provide a simple and easy way to eliminate some of the disadvantages of acrylic films while maintaining high strain performance.

Much more promising results have been obtained in prestrain-locked VHB-based IPNs [197–201]. These IPNs are fabricated by first prestraining the VHB film to very high strains and spraying a multifunctional monomer onto the film along with an initiator, then allowing the monomer and initiator to diffuse into the film. The additive monomers are then polymerized and form a second network elastomer within the VHB elastomer host. Upon releasing the IPN film from its support, the additive network will resist the contraction of the VHB host, preventing it from returning to its unstrained state. The IPN film will thus remain in a state wherein the VHB film is locked in tension and the additive network is locked in compression. The process is outlined in Fig. 1.14.

The resulting films are capable of performance similar to prestrained VHB films without the need for a support frame. Ha et al. have reported results for IPNs incorporating bifunctional 1,6-hexandiol diacrylate (HDDA) [197, 198] and trifunctional trimethylolpropane trimethacrylate (TMPTMA) monomers [199, 200]. With no externally applied prestrain, these prestrain-locked IPN films have matched the performance of highly prestrained ($300 \times 300\%$ biaxial prestrain) VHB4910 acrylic elastomers in terms of strain, electromechanical coupling factor, and energy density [201]. With no externally applied prestrain, the TMPTMA-based IPN films are capable of thickness strains as high as -75%, with an energy density of 3.5 MJ m^{-3}, pressure of 5.1 MPa, and coupling efficiency of 94%, with a breakdown field of 420 MV m^{-1}. Figure 1.15 shows the actuation of VHB-based IPN films with HDDA and TMPTMA additives.

IPN films have the added benefits of reducing viscoelasticity and enhancing mechanical stability as compared to regular VHB acrylic elastomers [201]. Reduced viscoelasticity has led to improvements in mechanical efficiency as seen in Fig. 1.16. These materials should open the door to a host of new actuator configurations and applications with minimal supporting structures and very high power-to-mass ratios.

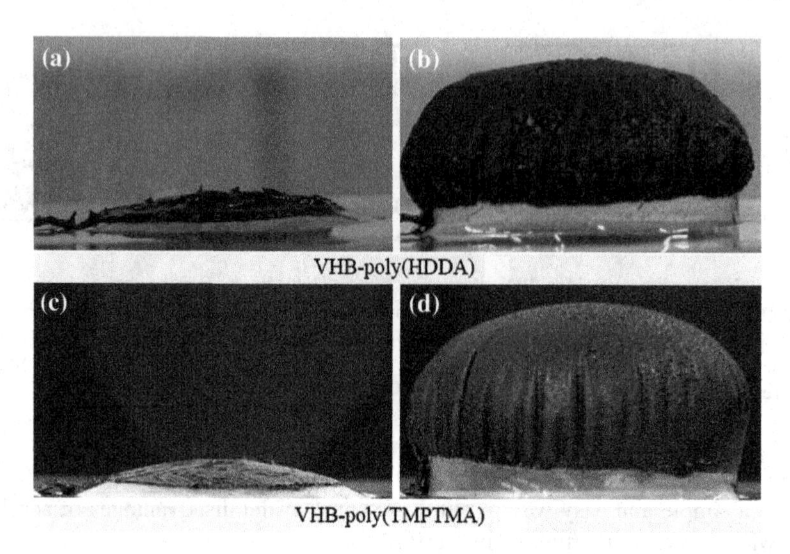

Fig. 1.15 VHB-based IPN films before and after actuation for films with HDDA (a to b) and TMPTMA (c to d) additives in a diaphragm configuration with no externally applied prestrain. Only a small bias pressure was used in the diaphragm chamber as evidenced by the small bulge in the film prior to actuation

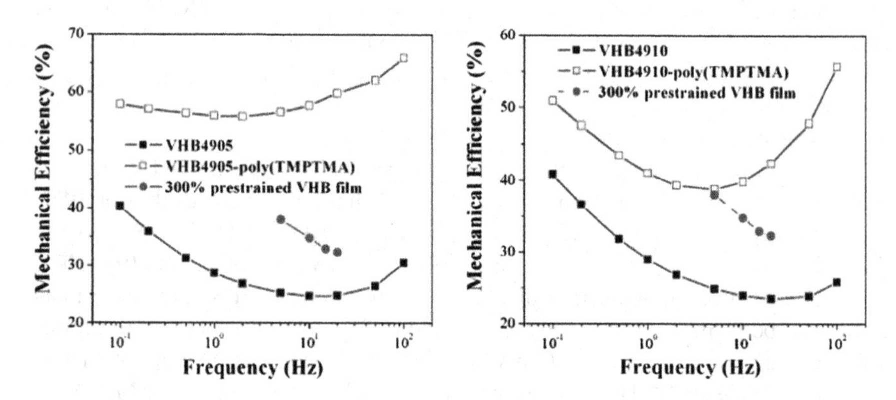

Fig. 1.16 Improved mechanical efficiency for VHB-based IPN films over neat VHB films in both the highly prestrained and non-prestrained states. The improved efficiency over prestrained VHB acrylic elastomers is attributed to a reduction in viscoelasticity [201]. Proc SPIE 2008, reprinted with permission

1.4.4 Thermoplastic Block Copolymers

Thermoplastic block copolymeric elastomers are also of interest as dielectric elastomer materials. These polymers differ from conventional elastomers in that they possess physical crosslinks rather than chemical ones. In these polymers

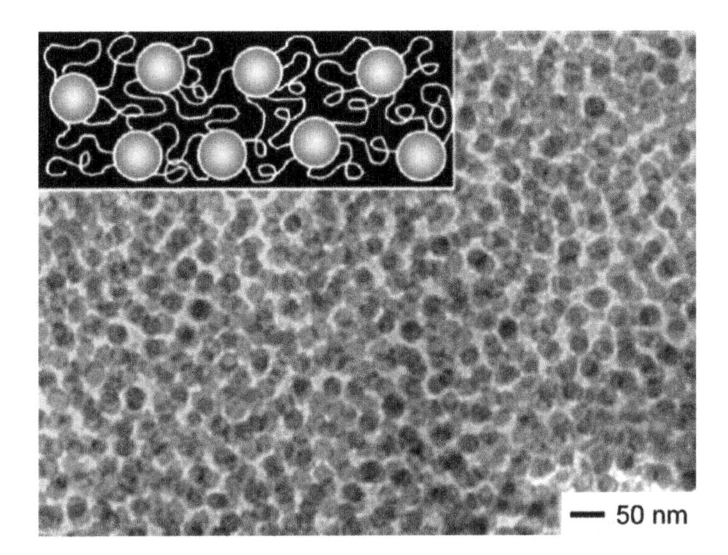

Fig. 1.17 TEM image showing the presence of glassy micelles of a SEBS copolymer in midblock sensitive oil with a polymer loading of 10 wt%. The inset is a depiction of the glassy micelle-stabilized polymer network [24]. Soft Matter 2007, reprinted with permission

flexible elastic chain segments separate rigid segments. The rigid sections act as the binding points between chains, while the flexible segments allow for large deformations.

Recently, Shankar et al. reported on nanostructured polystyrene-block-poly(ethylene-co-butylene)-block-polystyrene (SEBS) triblock copolymers swollen with a midblock sensitive oil [202, 203]. At relatively high oil concentrations, the thermoplastic elastomer behaves as a physical network where glassy styrene micelles serve as thermally reversible crosslinks. The glassy micelles are linked by rubbery ethylene-co-butylene midblocks swollen in the oil, giving the copolymer its elastomeric properties. Figure 1.17 shows a TEM image of a SEBS copolymer in midblock sensitive oligomeric oil with a concentration of 10 weight % polymer; the inset is a depiction of the network linked by the glassy micelles. Actuation strains as high as 245% in area (71% thickness) were reported in highly prestrained actuators, exceeding the maximum reported values for silicone and rivaling those reported for VHB acrylic elastomers.

Because of the ability to tune the composition by varying the copolymer molecular weight and the weight fraction of the polymer and oil, materials can be fabricated with tensile moduli ranging from 2 to 163 kPa with actuation strains being the highest for the low modulus variety and decreasing to 30% in area (22% thickness) at the high modulus range. Coupling efficiencies achieved were as high as 92% for low polymer loadings and lower molecular weight (161 kg mol^{-1}) and reduced to approximately 40% at high polymer loadings and higher molecular weight (217 kg mol^{-1}). A maximum energy density of 289 kJ m^{-3} was achieved at intermediate polymer weight fractions for the higher molecular weight

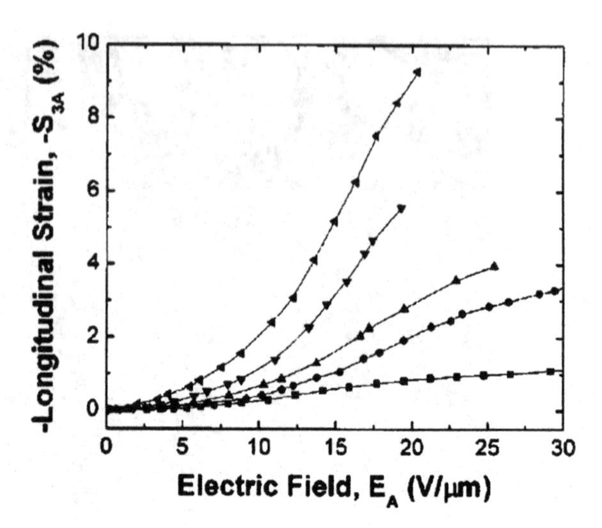

Fig. 1.18 Thickness strain S_{3A} as a function of the applied-field amplitude for composites of PANI/yPolyCuPc/PU (from *lowest strain* to *highest strain*): 0/0/100, 0/15/85, 4.6/15/85, 9.3/15/85, 14/15/85 [184]. Appl Phys Lett 2004, reprinted with permission

copolymer. A recent report showed that the materials are capable of blocking stresses as high as 442 kPa with a breakdown field of 203 MV m^{-1} at a polymer loading of 30 weight % [204]. The material also exhibited low cyclic hysteresis. These nanostructured polymers also display favorable actuation characteristics in their non-prestrained state when compared with non-prestrained VHB acrylic elastomers [203]. Unfortunately it appears as though large strains require low polymer loadings, while high blocking stress and breakdown field are limited to higher polymer loadings. The issue is the requirement for a midblock sensitive oil to allow for easier chain movement in the ethylene-co-butylene midblocks. Without a high concentration of solvent, the chain mobility is reduced and strains are limited; however, as the concentration is increased, the breakdown field suffers and leakage current may increase, limiting the blocking pressure and reducing efficiency. Depending on the solvent used, there may also be leaching issues that can act to degrade the polymer over time.

Polyurethane (PU) also falls under the class of thermoplastic polymers. The electromechanical response of polyurethane is due to both Maxwell pressure and electrostriction. The Maxwell pressure has been found to have a significant contribution to the strain response of polyurethane films above the glass transition temperature [205]. While polyurethane films initially proved promising, they fell to the sidelines soon after due to the dramatic improvements in the actuation properties of silicone and acrylic films with prestrain. The majority of research since then has focused on polyurethane based composites. Cameron et al. reported on their findings pertaining to graphite loaded polyurethane films [206]. They reported an increase in the actuation stress by a factor of over 500 and a relative permittivity beyond 4000 for graphite loadings near the percolation threshold. Unfortunately, such high loadings resulted in an increase in the dielectric loss factor by several orders of magnitude and an increase in modulus by a factor of 5. The result was a dramatic increase in the leakage current, power consumption, and a reduction in maximum actuation strain (due to the inability to reach high fields). Nam et al. demonstrated the ability to tune the bulk permittivity and ionic conductivity of polyurethane-montmorillonite (MMT) nanocomposites by varying the gallery heights of the (MMT) nanoplatelets through the addition of different counter ions [207]. Resulting permittivity values were shown to vary from below that of pristine polyurethane for both low and high gallery heights, to above the value for intermediate gallery heights, indicative of an optimal gallery height for such composites.

Huang et al. have demonstrated an all-organic, three-component polyurethane-based composite with a high electromechanical response [184]. By combining a high dielectric constant copper phthalocyanine oligomer (PolyCuPc) and conductive polyaniline (PANI) into a PU matrix, they have been able to achieve an electromechanical strain of 9.3% and elastic energy density of 0.4 J cm^{-3} under an electric field of only 20 MV m^{-1}. This approach overcomes issues associated with other percolative composites in which the loading of conductive filler must be near the percolation threshold in order to achieve meaningful increases in dielectric constant, which has a deleterious effect on the dielectric breakdown strength of the composite. In the composite, the PolyCuPc enhances the dielectric constant of the PU matrix; the two-component system also acts as the host for the conductive PANI that further enhances the dielectric response via a percolative phenomenon at much lower concentrations than in single component systems. Figure 1.18 shows the increase in thickness strain achieved in the three-component composite at different filler loadings.

Recent results have been reported on another polyurethane based composite film [208, 209]. A carbon nanopowder was added into a polyether-type thermoplastic (TPU5888 from Estane) at loadings up to 1.5 volume%. The maximum thickness strain of the composite was 12% versus 8% in the pure polymer, with a maximum pressure of 0.14 MPa (unchanged), and a response speed in the ms time range under a driving field of 8 MV m^{-1}. The results may have been limited by the use of sputtered gold electrodes that can contribute to device stiffness and lose conductivity at moderate strains. The energy density of the composite was estimated using FEM simulations to be 4.3 kJ m^{-3}, a factor of approximately 1.7 increase over the pure polymer.

P(VDF-TrFE), a well-studied electrostrictive polymer, is also thermoplastic. By dispersing copper-phthalocyanine (CuPc), a high dielectric constant metallorganic compound, into a P(VDF-TrFE) matrix, the resulting composites maintained the flexibility of the matrix with a very high dielectric constant (425 at 1 Hz) and relatively low dielectric loss [210]. The dielectric constant was shown to vary with both electric field and frequency and was highest at high fields and low frequencies. The strain exhibited a quadratic dependence on electric field, as expected, and a thickness strain of -1.91% was achieved at a field of 13 MV m^{-1} with an elastic energy density as high as 0.13 MJ m^{-3}. The strain response of the material was attributed to a number of mechanisms, including Maxwell stress and electrostriction.

1.4.5 Other Engineered Elastomers

It is interesting to note that the elastomers that provide the best actuation characteristics (VHB acrylics and silicones) were not designed for use as dielectric elastomers. Their noteworthy performance is not the result of targeted materials developed, but rather a fortuitous coincidence. The development of new engineered dielectric elastomers has made slow progress; however, in recent years several new developments have been presented that indicate that research is moving toward improved materials with the targeted application of dielectric elastomer actuators for artificial muscle applications.

Jung et al. have developed a synthetic elastomer composed of acrylonitrile butadiene rubber copolymer [211, 212]. The properties of the copolymer can be tuned by changing its composition. Reported data for dielectric constant, elastic modulus, and strain relaxation are promising (see Table 1.2). The synthetic elastomer provides some improvement over VHB and some silicone films under certain conditions; however, the tests were limited to low prestrain (60% radial), where the performance of VHB films is poor.

The same group has recently reported on the effects of a plasticizer (dioctyl phthalate (DOP)) and a high-K ceramic (TiO_2) on the actuation performance of their synthetic elastomer [213]. The addition of DOP showed the expected results of lower modulus, increased strain, and increased elastic energy density. The addition of TiO_2 had the effect of decreasing the modulus, increasing the dielectric constant, and increasing the strain and elastic energy density up to an optimal value, after which the values were seen to decrease. Maximum energy density of 1.2 kJ m^{-3} was achieved for a synthetic elastomer with a DOP loading of 100 parts per hundred rubber (phr) with a radial strain of 1.62%. A maximum strain of 3.04% was reported for a DOP loading of 80 phr and a TiO_2 loading of 30 phr. The maximum dielectric constant achieved was 11.1 for DOP and TiO_2 loadings of 80 and 40 phr respectively, but strain and elastic energy density were limited to 1.7% and 0.5 kJ m^{-3} respectively. The values, albeit lower than the peak performance of most conventional DE materials, were achieved at relatively low electric fields (20 MV m^{-1} or lower).

1.5 Compliant Electrode Materials

Identifying or developing a suitable compliant electrode is also a key factor in achieving good DE performance. A number of electrode materials have been explored. Original tests were performed using thin metal films. While these films provided good electrical conductivity, they limited strains to approximately 1%. Good compliant electrode materials should maintain high conductivity at large strains, have negligible stiffness, maintain good stability, and be fault tolerant. Common solutions include metallic paints (e.g. silver grease or paint), carbon grease, graphite, and carbon powder. Carbon grease electrodes are the most commonly used solution as they provide good conductivity even at very high strains, are cheap and easy to apply, and provide good adhesion to most DE materials while having minimal negative impacts on actuation performance. Dry graphite and carbon powder are also cheap, easy to apply, and have the additional benefit of being easy to handle. These dry electrode materials are better suited to multilayer devices where carbon grease electrodes result in slippage between adjacent layers which can eventually result in inhomogeneities in the electrode coverage; however, they tend to lose conductivity at high strains as the individual particles are pulled apart and lose contact. A comparative evaluation of some of the early electrode materials was performed by Carpi et al. in 2003 [214].

Improved metal electrodes have been developed by Benslimane et al. [215]. The electrodes are capable of achieving an anisotropic planar stain of 33% before losing electrical contact. The key to their design was patterning the surface of the elastomer itself prior to depositing a layer of silver. The films are able to expand in the corrugated direction, while expansion in the lateral direction is inhibited. Similarly, Lacour et al. achieved 22% strain with gold electrodes by first applying a compressive stress to the elastomer film in order to create surface waves [216].

Recent studies have been performed on alternative electrode materials. Nanosonic has developed low modulus, highly conducting thin film electrodes by molecular level self-assembly processing methods capable of maintaining conductivity up to strains of 100% [217, 218]. Recent developments have enabled the reduction of the modulus to less than 1 MPa and an increase in the strain to rupture to 1000% [219]. A version of the material is commercially available under the name Metal Rubber™. Delille et al. have developed novel compliant electrodes based on a platinum salt reduction [220]. The platinum salt is dispersed into a host elastomer and immersed in a reducing agent. A maximum conductivity of 1 S cm^{-1} was observed and conductivity was maintained for strains up to 40%.

In order to fully exploit the scalability of DE actuators, it necessary to be able to pattern electrodes on the micro scale as well. Rosset el al have explored the use of ion implanted metal electrodes in PDMS [221–224]. Their results show that conductivity can be maintained for strains up to 175% and can remain conductive over 10^5 cycles at 30% strain. This is of particular importance for MEMS microfluidic devices where the DE micro-actuators could be used as micro-pumps. The ion-implanted films maintained high breakdown fields (>100 MV m^{-1}) while the Young's modulus increased by 50–200% depending on the dose.

Studies performed by Yuan et al. on conductive polyaniline (PANI) nanofibers, P3DOT, and CNT thin films show that all three are capable of forming highly compliant electrodes with fault tolerant behavior [225]. Nanowires and tubes are of particular interest since they are capable of maintaining a percolation network at large strains, thus reducing the required electrode thickness while still allowing for maximum strain performance.

Further investigations by Lam et al. showed that PANI nanofibers films provided good actuation characteristics, provided fault tolerance, and had a negligible influence on the mechanical properties of the film, but lost conductivity over time

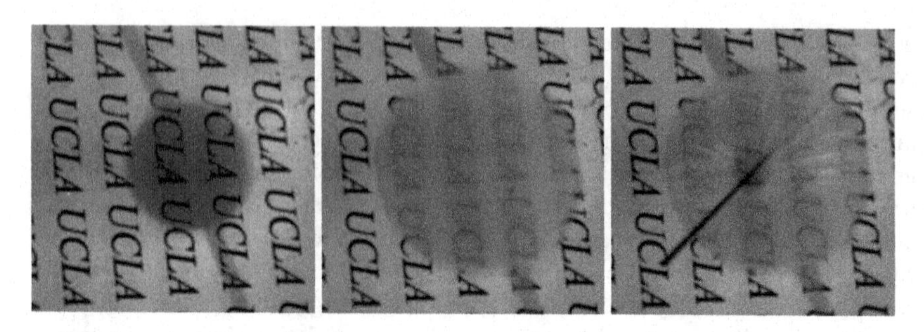

Fig. 1.19 Fault-tolerance of CNT electrodes. DEAs with CNT electrodes are able to withstand puncture and maintain a high level of strain due to self-clearing of CNT around puncture. From *left* to *right*: prestrained VHB acrylic actuator with CNT electrodes; actuated; actuated with a cactus pin through the active area

[226]. When tested on VHB 4905 films prestrained biaxially by 300 × 300%, the electrodes provided an maximum area strain of 97% at 3 kV and demonstrated self-clearing with a preserved strain of 91% after the first clearing event.

More recent results on CNTs were much more promising [227, 228]. Two types of CNT thin films were tested: functionalized P3-single walled nanotubes (SWNTs) and raw (non-functionalized) SWNTs. Both films provided excellent actuation characteristics (on par with carbon grease), had a negligible influence on mechanical properties of the film, and remained stable over longer periods of time. In addition, the carbon nanotube films could be made thin enough to remain optically transparent for use as transparent compliant electrodes [229].

The ultrathin PANI and CNT thin films are capable of "self-clearing", a process wherein the electrode material is burnt off locally in the event of an electrical short through the dielectric film [225–228]. Dielectric failure is one of the leading causes for premature device failure of DEs and results in terminal failure in carbon grease, graphite, and carbon powder electroded devices. The fault tolerance of devices with CNT electrodes is demonstrated in Fig. 1.19; in the tests a prestrained circular VHB acrylic actuator was punctured with a cactus pin and maintained very high actuation strains. In separate tests, the actuators were driven at a high field until the occurrence of a localized breakdown, the actuators "self-cleared" and recovered well from the local failures and retained high actuation strains. An SEM image of a "self-cleared" region is shown in Fig. 1.20. The image shows that the CNTs in the region surrounding the short have been burnt off. The addition of self-clearing introduces fault tolerance to DEs and can dramatically increase device lifetime. Results for CNT-based electrodes have demonstrated increases in constant actuation lifetime over carbon grease electrode based devices by two orders of magnitude; and demonstrated the ability to withstand and recover from several localized dielectric breakdown events.

Further increases in operational lifetime have been achieved through the addition of a thin layer of dielectric oil over the CNT electrodes [230]. Due to the high field amplification at the CNT tips, corona discharging through the air is an

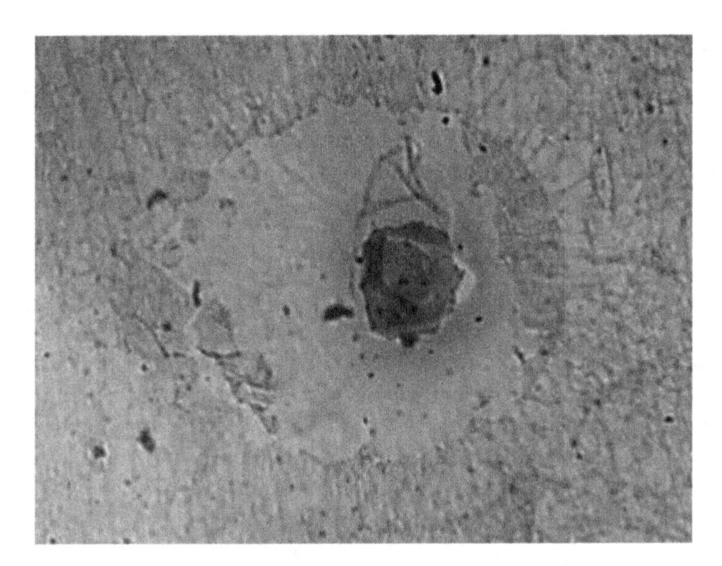

Fig. 1.20 Cleared area on a CNT electrode. Localized dielectric breakdown results in an electrical short through the film. Corona discharging burns away the CNTs in the area surrounding the short, isolating it from the rest of the electrode and allowing the device to continue operating

issue. In P3-SWNT films, this results in a slow degradation of the electrode conductivity resulting in lower actuation strains over time. Raw-SWNT films suffer from unabated breakdown-clearing events, which persist until the film loses mechanical stability and fails. The dielectric oil fills in the gaps between adjacent nanotube tips and flows into the voids left behind after the clearing events, turning the corona discharging process into silent discharge. This results in fewer breakdown-clearing events and can prolong constant actuation life at high strains by over an order of magnitude.

With the improved fault tolerance and lifetime afforded by compliant carbon nanotube electrodes with dielectric oil, DE artificial muscle devices that are capable of providing high performance actuation reliably over an extended number of cycles should be possible and should push DE actuators and artificial muscles closer to commercialization.

1.6 DE Actuator Configurations

Because DEs are fabricated from conformable elastomers, they can be shaped into many actuator configurations over a wide range of dimensions. Most actuator designs use the area expansion of the DE film for actuation; however, multilayer stacked actuators exist wherein actuation is through a reduction in thickness. Typical designs incorporate support structures to maintain prestrain in the films, though materials and processing advances have allowed for frameless designs as well. The majority of the actuator configurations in use today were developed by SRI in the late 1990 s and early 2000s [157, 164, 231, 232]. These include rolled (spring and core free), tube, unimorph, bimorph, stretched-frame, diaphragm, bowtie, spider, and extender. Carpi et al. provide a detailed account of dielectric elastomer technology with an emphasis on actuator configurations and applications [233].

A number of other actuator configurations have been developed, including a bidirectional framed actuator [234], a multiple-degree-of-freedom double diaphragm-type actuator [235], a tube-spring actuator (TSA) [236], a reinforced cylindrical actuator [237], an active shell-based actuator [238], a cone actuator [239], a compound structure frame actuator [240], an inflated bending actuator [241], and thickness mode actuators that can be used in programmable deformable surfaces [242]. Additional efforts have been put into characterizing and modeling the performance of different actuator configurations [243].

Yet another interesting DE actuator configuration, dubbed DEMES, which stands for Dielectric Elastomer Minimum Energy Structure, has been introduced by Kofod et al. [244, 245] and further developed by O'Brien et al. [246]. This type of actuator relies on a thin bendable frame. The frames are designed such that, when a prestrained DE layer is fixed to the frame, the forces exerted by the prestrained DE layer will cause the frame to curl until the forces balance; the result is that the device is curved in the equilibrium rest state. Upon actuation the DE layer will relax causing the actuator to uncurl until it reaches the equilibrium actuated state, which will vary with the applied field.

Plante et al. proposed using bistable binary actuator systems [247]. These actuators are capable of transitioning between two stable rest states and thus do need to be actuated continuously for extended periods of time, which can dramatically improve overall lifetime.

Artificial Muscle Incorporated (AMI) has developed a Universal Muscle Actuator (UMA) that has been proposed for use as an auto-focusing lens for camera phones [248]. This design relies on coupled prestrained circular framed actuators linked antagonistically and mechanically biased; they are spaced apart around the outer edge and connected at the center. Actuation of one of the films allows the other to relax, resulting in out-of-plane linear displacements.

From the perspective of artificial muscle design, the two most interesting device designs are the spring roll and stacked actuator. Both of them are able to effectively couple the deformations of DEs to provide linear actuation. Spring rolls are interesting from the perspective that they are capable of providing a combination of large linear strains ($\sim$30% of the active area) with relatively large output forces ($\sim$21 N for a cylindrical device measuring 65 mm in length and 12 mm in diameter) [249, 250]. The main drawback of spring roll actuators is that their design requires several passive or semi-active components such as a compression spring and end caps. A spring roll actuator can be seen in Fig. 1.21 along with a schematic of the components. A DE/electrode/DE/electrode layered film strip is rolled around a spring compressed between two end caps. When released, the

Fig. 1.21 Spring roll DEA. Rolls are fabricated by compressing a spring between two end caps and rolling a DE/electrode/DE/electrode layered strip around it. Increasing the number of layers wrapped around the spring core increases the output force [251]. Proc SPIE 2006, reprinted with permission

spring acts to maintain the prestrain in the film and prevent buckling. Reliability is an issue as the spring and end caps may generate areas of localized high stress that may result in premature failure.

Spring rolls capable of both bending and elongation have been reported by Pei et al. [251–253]. Bending is achieved by patterning the electrodes; for two and three degree-of-freedom actuators, two and three electrically isolated electrode pairs are required respectively. By actuating one electrode pair, only a portion of the actuator will elongate, causing the actuator to bend. These actuators are capable of bending angles as high as 90 degrees with lateral forces higher than 1.5 N for cylindrical devices measuring 9 cm in length and 2.3 cm in diameter. The ability to both bend and elongate opens up a number of new possibilities for multiple-degree-of-freedom spring rolls. These will be explored in the next section.

Stacked and folded configurations consist of tens to thousands of DE films stacked together and utilize the reduction in thickness rather than the area expansion of the DE film as the means of actuation. Employing a large number of layers amplifies the displacements. Several configurations have been proposed and tested [254–259]. The simplest device consists of laminated layers of DEs sandwiching compliant electrodes. Several silicone-based actuators have been developed and have demonstrated linear strains in excess of 15%.

To reduce manufacturing difficulties and the likelihood of electrical shorts that may result from the presence of disjointed electrodes, a helical design was proposed where only two continuous electrodes were required [255–256]. The actuator was capable of a -5% compressive strain at a field of only 14 MV m^{-1} using a softened silicone. A third configuration has also been proposed wherein a single long strip of DE film with electrodes on either side is simply folded in such a way as to keep the opposing electrodes isolated [257–259]. This configuration is perhaps the easiest to fabricate and yielded compressive strains and stresses up to 15% and $\sim$6 kPa respectively, but may result in non-uniform displacements. The three stacked actuator configurations are shown in Fig. 1.22.

Another method to reduce manufacturing difficulties was developed by Schlaak et al. [260, 261]. They implemented a process wherein the components of a

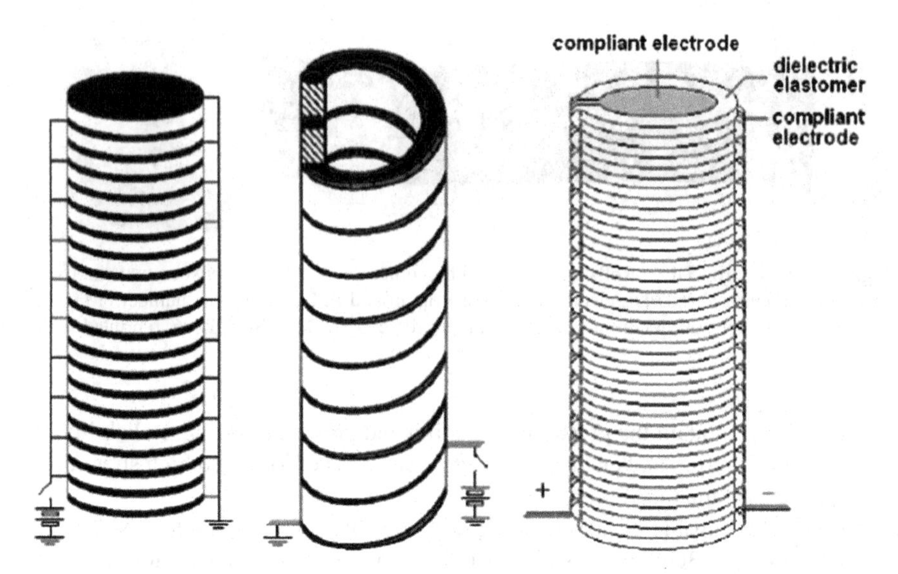

Fig. 1.22 Stacked linear contractile actuator configurations. From *left* to *right*: *Stacked* device wherein alternating layers of elastomer and electrode are stacked together; *helical* device where two complementary *helical* elastomer strips and electrodes are interlocked; *folded* device where a single strip of elastomer with electrodes on the *top* and *bottom* is folded upon itself [258]. Proc SPIE 2007, reprinted with permission

thermally cured silicone elastomer were spin-coated and thermally cured while the electrodes were spray coated using a contact mask.

Recent efforts by Arora et al. have led to the development of DE based fiber actuators [262]. These actuators are essentially miniaturized tube actuators. In their work, prestrain was applied by both uniaxial elongation as well as inflation. These actuators may be of interest for artificial muscle applications since they mimic the fibular nature of natural muscle; however, unlike natural muscle, their method of actuation is through elongation and the maximum strains reported were limited to 7% in the axial direction. Cameron et al. have performed similar work on co-extruded tube DE actuators [263]. Their co-extrusion process presents a fast, easy, scalable method to produce small diameter tubes, or hollow fibers, for use as DE actuators that would make these actuators very amenable to mass manufacturing. Strains, thus far, are limited to approximately 2%.

The use of IPN films, described in the Dielectric Elastomer Materials section, has enabled higher energy density actuators. VHB-based IPN films seem particularly suited to the stacked actuator configuration as they do not require prestraining. Kovacs and During have recently reported on the fabrication of such an actuator [264]. They constructed two prototypes: one consisting of 280 layers with an active diameter of 16 mm, a total diameter of 18 mm, and a height of 18.3 mm; the second consisting of 330 layers, an active diameter of 16 mm, a total diameter of 20 mm, and a height of 21.2 mm. The actuators achieved 46 and 35%

Fig. 1.23 Arm wrestling robot using dielectric elastomer *spring roll* actuators. *Left*: actuator assemblies used in the device. *Right*: demonstration of the arm wrestling robot [267]. Smart Mater Struct 2007, reprinted with permission

contractile strain respectively with both ends free, and 30 and 20% respectively with their ends fixed to rigid supports. The reduced strain performance for devices with fixed ends was attributed to the physical constraints imposed by the rigid end pieces. The results were obtained at voltages just above 4 kV. Maximum forces achieved were in excess of 30 N. IPN films have also been incorporated in core free rolled actuators [265]. Since the film is free standing, no spring core is needed to maintain prestrain in the film. As a result, the shelf and actuation lifetime, as well as the specific volume energy density of the device, have been improved.

1.7 DE Applications

1.7.1 DE Actuator Applications

When placed in an antagonistic arrangement, linear DE actuators have demonstrated the ability to mimic the motion of natural muscle. We are still some way off reliable reproduction of natural muscle's performance; however, these preliminary results show a great deal of promise. When coupled with new advances in artificial skin and other related technologies that may reduce the performance requirements for artificial muscles, advances in DE actuators may allow for commercial applications sooner than would otherwise be anticipated [266].

Perhaps the best example of both the promise that DE actuators and other EAP technologies show, and hurdles they must overcome, was the 2005 EAP arm wrestling competition organized by Bar-Cohen. Kovacs et al. reported on the construction of one such robot as seen in Fig. 1.23 [267].

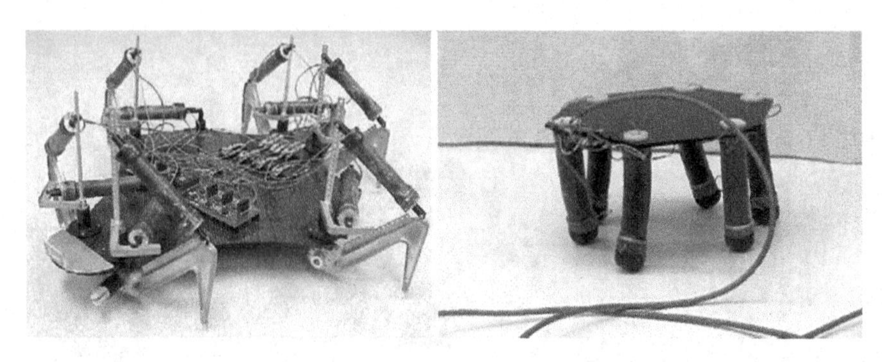

Fig. 1.24 FLEX II and MERbot walking robots. FLEX II uses two dielectric elastomer *spring roll* actuators per leg while MERbot uses multiple-degree-of-freedom *spring roll* actuators [252, 275]

Other artificial muscle applications have been demonstrated as well. Carpi et al. used helical contractile linear actuators and buckling actuators to actuate eyeballs for use in an android face [258, 268, 269]. Another eyeball actuator has been developed by Liu et al. based on their inflated actuator design; their actuator is capable of generating eyeball rotations from -50 to 50° [241]. Kornbluh et al. have also reported on a mouth driven by a DE actuator [4]. Biddis and Chau have provided a good review on the challenges and opportunities of DE actuators for upper limb prosthetics [270].

DE actuators have been used in a number of biomimetic robots ranging from simple hopping robots [271] and inchworm robots [4, 272–275] to more complex walking robots, flapping-wing robots, and serpentine or octopus arm-inspired actuators [4, 238–240, 275, 276]. Two particularly interesting walking robots developed by SRI, dubbed Flex II and MERbot, can be seen in Fig. 1.24. Utilizing spring roll actuators, Flex II was capable of a respectable 3.5 cm s^{-1} [275]. MERbot used multiple-degree-of-freedom spring rolls to achieve speeds in excess of 13 cm s^{-1} [252].

Potential applications of DE actuators are by no means limited to artificial muscles. A plethora of other DE applications have been proposed and demonstrated that can potentially be used in humanoid devices. These include loudspeakers [277–279], variable diffraction gratings [278], tunable transmission gratings [280], and micro-optical zoom lenses [281], among others. These applications may find use in biomimetics as a method to bestow polymer robots with the abilities to speak and to focus their vision.

Outside of biomimetics, dielectric elastomers may still be of use in other human-related applications including refreshable Braille devices [260, 282–284], hand rehabilitation splints [285], MRI compatible machines [247, 286–288], micro-fluidic devices [289, 290], force feedback [251], and wearable tactile interfaces [284, 291]. Other potential applications include actuators for lighter-than-air vehicles [292, 293], tunable phononic crystals [294–296], and variable phase retarders [297]. Beck et al. have also developed a sub-100 nm resolution total internal reflection fluorescence microscope using a tunable transmission grating and transparent phase shifters actuated by electroactive polymers [298].

1.7.2 Extending Applications of Dielectric Elastomers

Since DEs behave essentially as variable capacitors, they may also be used as sensors and even as generators. Pelrine et al. introduced the concept of DE generators in 2001 [299]. Several generator configurations have since been conceived [4, 278, 299–305]. Conceptually, dielectric elastomer devices can convert mechanical deformations into electrical energy based on the difference in capacitance between the device in the stretched and contracted states. If a bias voltage is applied across the electrodes in the stretched state with capacitance C_s, a certain amount of charge, Q, will be stored on the film at a potential V_b. If the film is then allowed to contract, the capacitance will decrease to a value C_c since the electrodes reduce in area and increase in separation. If the leakage current is negligible and no charge is allowed to flow off the DE device, then the amount of charge, Q, will remain unchanged, but the potential of the charge will increase to $V_b + V_g$, where V_g is the electrical potential energy generated from the release of stored elastic energy in the elastomer. The generated energy, e_{gen}, can be estimated by taking the difference between the stored energy on the DE between the contracted and stretched states, e_c and e_s respectively:

$$e_{gen} = e_c - e_s = \frac{1}{2}\left(C_c V_c^2 - C_s V_c^2\right) = \frac{1}{2}C_c\left(V_g^2 - V_b V_g\right) \tag{1.11}$$

The generated energy for a particular material can be maximized by operating the DE at a high bias field and stretching the film to large strains. The maximal energy that can be converted, however, is limited by a number of phenomena that impact both the generator and actuator modes of operation. These include electrical breakdown, electromagnetic instability (pull-in), loss of tension, and rupture by stretch [306].

Jean-Mistral et al. have developed a more complete model of the generator mode, incorporating viscous and hyperelastic effects [307]. Their treatment takes into account viscous and electrical losses to produce more realistic estimates of the generated energy available to an external load. Ihlemfeld and Qu have recently reported that an EAP actuator model with full electrical–mechanical dynamics can be used as a generator model with the generator input force equivalent to the actuator disturbance force [308]. This is good news for the development of DE generators as the modeling work performed for actuator applications can be easily leveraged.

By monitoring the capacitance of the DE films, it is also possible to determine the level of strain in the elastomer [309]. In this manner, a DE device can also act as a strain or deflection sensor. Methods have also been developed that allow for transient strain analysis during actuation, allowing for the creation of "self-sensing" DE actuators with integrated extension sensors [310–313]. Chuc et al. have also developed a "self-sensing" DE actuator capable of measuring the force exerted upon it [314]. This "self-sensing" ability, while undergoing actuation, should make the development of controls systems easier and should allow for more compact device assemblies.

In certain configurations, DEs may also be operated as spring elements with variable stiffness and damping characteristics [4]. If an actuator is held at a certain displacement, changing the applied electric field will vary the Maxwell stress across the device and thus the mechanical impedance.

The benefit of being able to sense deflections, generate energy, and modulate stiffness may have substantial impacts on the use of dielectric elastomers as artificial muscles in robotic and prosthetic applications. The DE elements could act as artificial analogs of natural muscle, sensory systems, and digestive systems. A robot consisting of DE elements should therefore someday be capable of controlled motion without the need for additional sensors, and self-sustainability without requiring and an external source of electricity.

1.8 Concluding Remarks

Electroactive polymers represent an important category of responsive materials for the transduction between electrical and mechanical energies. There are a diverse range of EAPs with distinctive actuation mechanisms and passive structural properties. Many EAPs exhibit modest to large actuation strain, specific energy density, modest to high actuation stress, on the same order of magnitude of performance as natural muscle. Dielectric elastomers appear to best reproduce the multifunctionality of muscles with unprecedented large actuation strain, high stress, specific energy density, and mechanical compliance. The high performance dielectric elastomers so far are largely commercial elastomers produced for unrelated applications. Prestrain, IPN technology, or plasticizing additives further extend the breadth of the performance scope. There should be plenty of room for further improvement in the polymer materials development for high actuation speed, specific power density, and energy efficiency. The requirement for highly compliant electrodes is also essential, particularly with regard to patterning, fabrication complexity, and actuation stability. The available large strains have enabled a number of novel device configurations. This is an area where more exciting developments are expected as more engineers start to explore DE actuators. Commercial product developments are being pursued, though manufacturing yield and operational stability remain problematic. There are many more developers waiting for the driving voltage to come down, from kilovolts to the low hundreds of volts range. This could be achieved by either reducing the thickness of the extremely soft DE films to a few microns, which is a paramount manufacturing challenge, or by increasing the dielectric constant by two orders of magnitude, a goal that is currently being researched. In our opinion, the kV driving voltage is manageable as long as the overall power and stored energy are kept low. With proper insulation, DE actuators should be applicable to replace or augment muscle functions in biomimetic robots, prosthetics, implanted devices, and various controls.

Acknowledgments We would like to thank the National Science Foundation NSF Award # CBET-0933556 for financial support.

References

1. Bar-Cohen Y (2004) Electroactive polymer (EAP) actuators as artificial muscle, 2nd edn. SPIE Press, Bellingham
2. Madden JD, Vandesteeg NA, Anquetil PA, Madden PG, Takshi A, Pytel AZ, Lafontaine SR, Wieringa PA, Hunter W (2004) Artificial muscle technology: physical principles and naval prospects. IEEE J Oceanic Eng 29:706
3. Pelrine R, Kornbluh R, Joseph J, Heydt R, Pei Q, Chiba S (2000) High-field deformation of elastomeric dielectrics for actuators. Mater Sci and Eng C 11:89
4. Kornbluh R, Pelrine R, Pei Q, Heydt R, Stanford S, Oh S, Eckerle J (2002) Electroelastomers: applications of dielectric elastomer transducers for actuation, generation, and smart structures. Proc SPIE EAPAD 4698:254
5. Hollerbach J, Hunter I, Ballantyne J (1992) In: Khatib O, Craig J, Lozano-Perez T (eds) The robotics review, vol 2. MIT Press, Cambridge 2:299–342
6. Honda Motor Company, http://world.honda.com/ASIMO/
7. Mirfakhrai T, Madden JD, Baughman R (2007) Polymer artificial muscles. Mater Today 10:30
8. Dickinson MH, Farley CT, Full RJ, Koehl MAR, Kram R, Lehman S (2000) How animals move: an integrative view. Science 288:100
9. Hunter I, Lafontaine S (1992) A comparison of muscle with artificial actuators. Tech. Dig. IEEE Solid State Sensors Actuators Workshop. 178–185
10. Bar-Cohen Y (2002) Electro-active polymers: current capabilities and challenges. Proc SPIE 4695:1
11. Bar-Cohen Y (2004) EAP as artificial muscles: progress and challenges. Proc SPIE 5385:10
12. Roentgen WC (1880) About the changes in shape and volume of dielectrics caused by electricity. In: Wiedemann G (ed) Annual Physics and Chemistry Series 11(III). JA Barth, Leipzig, Germany pp 771–786 (in German)
13. Ashley S (2003) Artificial muscles. Sci Amer 10:52–59
14. Baughman R (2005) Playing nature's game with artificial muscles. Science 308:63
15. Meijer K, Rosenthal M, Full RJ (2001) Muscle-like actuators? A comparison between three electroactive polymers. Proc SPIE 4329:7
16. Otero TF, Cascales JL, Fernandez-Romero AJ (2007) Attempting a classification for electrical polymeric actuators. Proc SPIE 6524:65240L
17. Trivedi D, Rahn CD, Kier WM, Walker ID (2008) Soft robotics: biological inspiration, state of the art, and future research. Appl Bionics Biomechanics 5:99
18. Cheng Z, Zhang Q (2008) Field-activated electroactive polymers. MRS Bull 33:183
19. Vinogradov AM (2008) Accomplishments and future trends in the field of electroactive polymers. Proc SPIE 6927:69270M
20. Pons JL (2005) Emerging actuator technologies: a micromechatronic approach. Wiley, New Jersey

21. Huber JE, Fleck NA, Ashby MF (1997) The selection of mechanical actuators based on performance indices. Proc R Soc Lond Ser A 453:2185
22. Wax SG, Sands RR (1999) Electroactive polymer actuators and devices. Proc SPIE 3669:2
23. Madden JD (2007) Mobile robots: motor challenges and materials solutions. Science 318:1094
24. Shankar R, Ghosh TK, Spontak R (2007) Dielectric elastomers as next-generation polymeric actuators. Soft Matter 3:1116
25. O'Halloran A, O'Malley F, McHugh P (2008) A review on dielectric elastomer actuators, technology, applications, and challenges. J Appl Phys 104:071101
26. Kim KJ, Tadokoro S (2007) Electroactive polymers for robotic applications: artificial muscles and sensors. Springer, London
27. Shahinpoor M, Kim KJ, Mojarrad M (2007) Artificial muscles: applications of advanced polymeric nanocomposites. CRC Press Taylor & Francis Group, Boca Raton
28. Carpi F, Smela E (2009) Biomedical applications of electroactive polymer actuators. Wiley, Chichester
29. Shahinpoor M, Kim KJ (2001) Ionic polymer-metal composites: I. Fundamentals. Smart Mater Struct 10:819
30. Kim KJ, Shahinpoor M (2002) A novel method of manufacturing three-dimensional ionic polymer-metal composites (IPMCs) bomimetic sensors, actuators and artificial muscles. Polymer 43:797
31. Shahinpoor M (2003) Mechanoelectrical phenomena in ionic polymers. Math Mech Solids 8:281
32. Wang XL, Oh IK, Lu J, Ju J, Lee S (2007) Biomimetic electro-active polymer based on sulfonated poly (styrene-bethylene-co-butylene-b-styrene). Mater Lett 61:5117
33. Nemat-Nasser S (2002) Micromechanics of actuation of ionic polymer-metal composites. J Appl Phys 92:2899
34. Shahinpoor M, Kim KJ (2005) Ionic polymer-metal composites: IV. Industrial and medical applications. Smart Mater Struct 14:197
35. Kaneto K, Kaneko M, Min Y, MacDiarmid AG (1995) Artificial muscle–electromechanical actuators using polyaniline films. Synth Met 71:2211
36. Akle BJ, Bennett MD, Leo DJ (2006) High-strain ionomeric-ionic liquid electroactive actuators. Sens Actuators A 126:173
37. Nemat-Nasser S, Wu YX (2003) Comparative experimental study of ionic polymer-metal composites with different backbone ionomers and in various cation forms. J Appl Phys 93:5255
38. Kim KJ, Shahinpoor M (2003) Ionic polymer-metal composites: II. Manufacturing techniques. Smart Mater Struct 12:65
39. Eamax Co., Japan, http://www.eamex.co.jp/
40. Kuhn W, Hargitay B, Katchalsky A, Eisenberg H (1955) Reversible dilation and contraction by changing the state of ionization of high polymer acid networks. Nature (London) 165:514
41. Katchalsky A, Zwick M (1955) Mechanochemistry and ion exchange. J Polym Sci 16:221
42. Fragala A, Enos J, Laconti A, Boyack J (1972) Electrochemical activation of a synthetic artificial muscle membrane. Electrochim Acta 17:1507
43. Osada Y (1987) Conversion of chemical into mechanical energy by synthetic-polymers (chemomechanical systems). Adv Polym Sci 82:1
44. Tanaka T (1979) Phase-transitions in gels and a single polymer. Polymer 20:1404
45. Osada Y, Hasebe M (1985) Electrically activated mechanochemical devices using polyelectrolyte gels. Chem Lett 9:1285
46. Shiga T (1997) Deformation and viscoelastic behavior of polymer gels in electric fields. Adv Polym Sci 134:131
47. Calvert P, Liu Z (1998) Freeform fabrication of hydrogels. Acta Metall Mater 46:2565
48. Liu ZS, Calvert P (2000) Multilayer hydrogels as muscle-like actuators. Adv Mater 12:288
49. Tondu B, Emirkhanian R, Mathé S, Ricard A (2009) A pH-activated artificial muscle using the McKibben-type braided structure. Sens Actuators A 150:124

50. Iijima S (1991) Helical microtubules of graphitic carbon. Nature 354:56
51. Baughman RH, Zakhidov AA, de Heer WA (2002) Carbon nanotubes–the route toward applications. Science 297:787
52. Baughman RH, Chanxing C, Zakhidov AA, Iqbal Z, Barisci JN, Spinks GM, Wallace GG, Mazzoldi A, De Rossi D, Rinzler AG, Jaschinski O, Roth S, Kertesz M (1999) Carbon nanotube actuators. Science 284:1340
53. Madden JDW, Barisci JN, Anquetil PA, Spinks GM, Wallace GG, Baughman RH, Hunter IW (2006) Fast carbon nanotube charging and actuation. Adv Mater 18:870
54. Mirfakhrai T, Madden JDW, Baughman RH (2006) Electrochemical actuation of carbon nanotube yarns. Smart Mater Struct 16:S243
55. Shin MK, Kim SI, Kim SJ, Kim SK, Lee H, Spinks GM (2006) Size-dependent elastic modulus of single electroactive polymer nanofibers. Appl Phys Lett 89:231929
56. Hughes M, Spinks GM (2005) Multiwalled carbon nanotube actuators. Adv Mater 17:443
57. Li YL, Kinloch IA, Windle AH (2004) Direct spinning of carbon nanotube fibers from chemical vapor deposition synthesis. Science 304:276
58. Baughman RH (2006) Materials synthesis–towering forests of nanotube trees. Nat Nanotechnol 1:94
59. Aliev AE, Oh J, Kozlov ME, Kuznetsov AA, Fang S, Fonseca AF, Ovalle R, Lima MD, Haque MH, Gartstein YN, Zhang M, Zakhidov AA, Baughman RH (2009) Giant-stroke, superelastic carbon nanotube aerogel muscles. Science 323:1575
60. Baughman RH, Shacklette LW, Elsebaumer RL, Plichta EJ, Becht C (1990) Conjugated polymeric materials: opportunities in electronics, optoelectronics, and molecular electronics. In: Brédas JL, Chance RR (eds), Kluwer, Dordrecht, pp 559–582
61. Baughman RH, Shacklette LW, Elsebaumer RL, Plichta EJ, Becht C (1991) Topics in molecular organization and engineering: molecular electronics, vol 7, Lazarev PI (ed), Kluwer, Dordrecht, 7:267–289
62. Baughman RH (1991) Conducting polymers in redox devices and intelligent materials systems. Makromol Chem Macromol Symp 51:193
63. Baughman RH (1996) Conducting polymer artificial muscles. Synth Met 78:339
64. Pei Q, Inganäs O (1992) Conjugated polymers and the bending cantilever method: electrical muscles and smart devices. Adv Mater 4:277
65. Smela E, Inganäs O, Lundström I (1995) Controlled folding of micrometer-size structures. Science 268:1735
66. Otero TF, Angulo E, Rodriguez J, Santamaria C (1992) Electrochemomechanical properties from a bilayer-polypyrrole nonconducting and flexible material artificial muscle. J Electroanal Chem 341:369
67. Herod TE, Schlenoff JB (1993) Doping-induced strain in polyaniline–stretch electrochemistry. Chem Mater 5:951
68. Pytel RS, Thomas E, Hunter I (2006) Anisotropy of electroactive strain in highly stretched polypyrrole actuators. Chem Mater 18:861
69. Pei Q, Inganäs O (1992) Electrochemical applications of the bending beam method .1. Mass-transport and volume changes in polypyrrole during redox. J Phys Chem 96:10507
70. Pei Q, Inganäs O (1993) Electrochemical applications of the bending beam method, a novel way to study ion-transport in electroactive polymers. Solid State Ion 60:161
71. Pei Q, Inganäs O (1993) Electrochemical applications of the bending beam method .2. Electroshrinking and slow relaxation in polypyrrole. J Phys Chem 97:6034
72. Pei Q, Inganäs O, Lundstrom I (1993) Bending bilayer strips built from polyaniline for artificial electrochemical muscles. Smart Mater Struct 2:1
73. Mazzoldi A, Della Santa A, De Rossi D (1999) Conducting polymer actuators: properties and modeling. In: Osada Y, De Rossi D (eds) Polymer sensors and actuators. Springer, Heidelberg, pp 207–244
74. Cole M, Madden JD (2005) The effect of temperature exposure on polypyrrole actuation. Mater Res Soc Symp Proc 889:W04

75. Smela E, Gadegaard N (1999) Surprising volume change in PPy(DBS): an atomic force microscopy study. Adv Mater 11:953
76. Hara S, Zama T, Takashima W, Kaneto K (2004) Artificial muscles based on polypyrrole actuators with large strain and stress induced electrically. Polymer J 36:151
77. Hara S, Zama T, Takashima W, Kaneto K (2006) Tris(trifluoromethylsulfonyl)methide-doped polypyrrole as a conducting polymer actuator with large electrochemical strain. Synth Met 156:351
78. Spinks GM, Mottaghitalab V, Bahrami-Samani M, Whitten PG, Wallace GG (2006) Carbon-nanotube-reinforced polyaniline fibers for high-strength artificial muscles. Adv Mater 18:637
79. Kaneto K, Kaneko M, Min Y, MacDiarmid AG (1995) Artificial muscle–electrochemical actuators using polyaniline films. Synth Met 71:2211
80. Lewis TW, Kane-Maguire LAP, Hutchinson AS, Spinks GM, Wallace GG (1999) Development of an all-polymer, axial force electrochemical actuator. Synth Met 102:1317
81. Spinks GM, Wallace GG, Ding J, Zhou D, Xi B, Scott TR, Truong VT (2003) Ionic liquids and polypyrrole helix tubes: bringing the electronic Braille screen closer to reality. Proc SPIE 5051:372
82. Pettersson PF, Jager EWH, Inganas O (2000) Surface micromachined polymer actuators as valves in PDMS microfluidic systems. 1st international IEEE-EMBS special topic conference on microtechnologies in medicine and biology. In: Dittmar A, Beebe D (eds) IEEE-EMBS, Lyon, France p 334
83. Della Santa A, Mazzoldi A, de Rossi D (1996) Steerable microcatheters actuated by embedded conducting polymer structures. J Intell Mater Sys Struct 7:292
84. Takase Y, Lee JW, Scheinbeim JI, Newman BA (1991) High-temperature characteristics of nylon-11 and nylon-7 piezoelectrics. Macromolecules 24:6644
85. Su J, Ma ZY, Scheinbeim JI, Newman BA (1995) Ferroelectric and piezoelectric properties of nylon 11/poly(vinylidene fluoride) bilaminate films. J Polym Sci B 33:85
86. Gao Q, Scheinbeim JI, Newman BA (2000) Dipolar intermolecular interactions, structural development, and electromechanical properties in ferroelectric polymer blends of nylon-11 and poly(vinylidene fluoride). Macromolecules 33:7564
87. Lovinger AJ, Davis GT, Furukawa T, Broadhurst MG (1982) Crystalline forms in a copolymer of vinylidene fluoride and trifluoroethylene (52/48 mol-percent). Macromolecules 15:323
88. Lovinger AJ (1982) Developments in crystalline polymers-1, Bassett DC (ed), Applied Science Publishers, London 1982:195
89. Lovinger AJ, Furukawa T (1983) Curie transitions in copolymers of vinylidene fluoride. Ferroelectrics 50:227
90. Xu Y (1991) Ferroelectric materials and their applications. North-Holland, Netherlands
91. Huang C, Klein R, Xia F, Li H, Zhang QM, Bauer F, Cheng ZY (2004) Poly(vinylidene fluoride-trifluoroethylene) based high performance electroactive polymers. IEEE Trans Dielectr Electr Insul 11:299
92. Tashiro KK, Takano M, Kobayashi Y, Chatani A, Tadokoro H (1984) Structural study on ferroelectric phase-transition of vinylidene fluoride-trifluoroethylene copolymers(III) dependence of transitional behavior of VDF molar content. Ferroelectrics 57:297
93. Zhang QM, Zhao J, Shrout T, Kim N, Cross LE, Amin A, Kulwicki BM (1995) Characteristics of the electromechanical response and polarization of electric field biased ferroelectrics. J Appl Phys 77:2549
94. Eguchi M (1925) On the permanent electret. Philos Mag 49:178
95. Sessler GM (ed) (1998) Electrets, 3rd edn. vol 1, Laplacian Press
96. Bauer S (2006) Piezeo-, pyro- and ferroelectrets: soft transducer materials for electromechanical energy conversion. IEEE Trans Dielectr Electr Insul 13:953
97. Bauer S, Gerhard-Multhaupt R, Sessler G (2004) Ferroelectrets: soft electroactive foams for transducers. Phys Today 57:37

98. Lovera D, Ruckdäschel H, Göldel A, Behrendt N, Frese T, Sandler JKW, Altstädt B, Giesa R, Schmidt HW (2007) Tailored polymer electrets based on poly(2,6-dimethyl-1,4-phenylene ether) and its blends with polystyrene. Eur Polym J 43:1195
99. Cheng ZY, Katiyar RS, Yao X, Bhalla AS (1998) Temperature dependence of the dielectric constant of relaxor ferroelectrics. Phys Rev B 57:8166
100. Cheng ZY, Zhang QM, Bateman FB (2002) Dielectric relaxation behavior and its relation to microstructure in relaxor ferroelectric polymers: high-energy electron irradiated poly(vinylidene fluoridetrifluoroethylene) copolymers. J Appl Phys 92:6749
101. Zhang QM, Bharti V, Zhao X (1998) Giant electrostriction and relaxor ferroelectric bahavior in electron-irradiated poly(vinylidene fluoride-trifluoroethylene) copolymer. Science 280:2101
102. Huang C, Klein R, Xia F, Li HF, Zhang QM, Bauer F, Cheng ZY (2004) Poly(vinylidene fluoride-trifluoroethylene) based high performance electroactive polymers. IEEE Trans Dielectr Electr Insul 11:299
103. Cheng ZY, Xu TB, Bharti V, Wang S, Zhang QM (1999) Transverse strain responses in the electrostrictive poly(vinylidene fluoride-trifluoroethylene) copolymer. Appl Phys Lett 74:1901
104. Guo S, Zhao XZ, Zhuo Q, Chan HLW, Choy CL (2004) High electrostriction and relaxor ferroelectric behavior in proton-irradiated poly(vinylidene fluoride-trifluoroethylene) copolymer. Appl Phys Lett 84:3349
105. Xia F, Cheng ZY, Xu H, Li H, Zhang QM, Kavarnos G, Ting R, Abdul-Sedat G, Belfield KD (2002) High electromechanical responses in a poly(vinylidene fluoride-trifluoroethylene-chlorofluoroethylene) terpolymer. Adv Mater 14:1574
106. Xu H, Cheng ZY, Olson D, Mai T, Zhang QM, Kavarnos G (2001) Ferroelectric and electromechanical properties of poly(vinylidene fluoride-trifluoroethylene-chlorotrifluoroethylene) terpolymer. Appl Phys Lett 78:2360
107. Garrett JT, Roland CM, Petchsuk A, Chung TC (2003) Electrostrictive behavior of poly(vinylidene fluoride-trifluoroethylene-chlorotrifluoroethylene). Appl Phys Lett 83:1190
108. Jayasuriya AC, Schirokauer A, Scheinbeim JI (2001) Crystal-structure dependence of electroactive properties in differently prepared poly(vinylidene fluoride/hexafluoropropylene) copolymer films. J Polym Sci Part B Polym Phys 39:2793
109. Wegener M, Hesse J, Richter K, Gerhard-Multhaupt R (2002) Ferroelectric polarization in stretched piezo- and pyroelectric poly(vinylidene fluoride-hexafluoropropylene) copolymer films .J Appl Phys 92:7442
110. Neese B, Wang Y, Chu BJ, Ren KL, Liu S, Zhang QM, Huang C, West J (2007) Piezoelectric responses in poly(vinylidene fluoride/hexafluoropropylene) copolymers. Appl Phys Lett 90:242917
111. Bao HM, Jia CL, Wang CC, Shen QD, Yang CZ, Zhang QM (2008) A type of poly(vinylidene fluoride-trifluoroethylene) copolymer exhibiting ferroelectric relaxor behavior at high temperature (similar to 100°C). Appl Phys Lett 92:042903
112. Xia F, Li H, Huang C, Huang MYM, Xu H, Bauer F, Cheng ZY, Zhang QM (2003) Poly(vinylidene-fluoride-trifluoroethylene) based high-performance electroactive polymers. Proc SPIE 5051:133
113. Su J, Hales K, Xu TB (2003) Composition and annealing effects on the response of electrostrictive graft elastomers. Proc SPIE 5051:191
114. Su J, Harrison JS, Clair TLS, Bar-Cohen Y, Leary S (1999) Electrostrictive graft elastomers and applications. Mater Res Soc Symp Proc 600:131
115. Warner M, Terentjev M (2003) Liquid crystal elastomers. Oxford Science Publications, Oxford
116. de Gennes PG, Chung TC, Petchsux A (1975) Réflexions sur un type de polymères nématiques. Seances Acad Sci Ser B 281:101
117. de Gennes PG (1997) A semi-fast artificial muscle. CR Acad Sci Paris Ser II B 324:343
118. Thomsen DL, Keller P, Naciri J, Pink R, Jeon H, Shenoy D, Ratna BR (2001) Liquid crystal elastomers with mechanical properties of a muscle. Macromolecules 34:5868

119. Wermter H, Finkelmann H (2001) Liquid crystalline elastomers as artificial muscles. e-Polymers 13:1, http://www.e-polymers.org
120. Li MH, Keller P (2006) Artificial muscles based on liquid crystal elastomers. Philos Trans R Soc A 364:2763
121. Lehmann W, Skupin H, Tolksdorf C, Gebhard E, Zentel R, Kruger P, Losche M, Kremer F (2001) Giant lateral electrostriction in ferroelectric liquid-crystalline elastomers. Nature 410:447
122. de Gennes PG (1982) Polymer liquid crystals: mechanical properties of nematic polymers. In: Cifferi A, Krigbaum WR, Meyer RB (eds) Academic Press, New York, pp 115–131
123. D'Allest JF, Maissa P, Ten Bosch A, Sixou P, Blumstein A, Blumstein RB, Teixeira J, Noirez L (1988) Experimental-evidence of chain extension at the transition-temperature of nematic polymer. Phys Rev Lett 61:2562
124. Li MH, Brûlet A, Davidson P, Keller P, Cotton JP (1993) Observation of hairpin defects in a nematic main-chain polyester. Phys Rev Lett 70:2297
125. Li MH, Brûlet A, Cotton JP, Davidson P, Strazielle C, Keller P (1994) Study of the chain conformation of thermotropic nematic main chain polyesters. J Phys II France 4:1843
126. Lehmann W, Hartmann L, Kremer F, Stein P, Finkelmann H, Kruth H, Diele S (1999) Direct and inverse electromechanical effect in ferroelectric liquid crystalline elastomers. J Appl Phys 86:1647
127. Leister N, Lehmann W, Weber U, Geschke D, Kremer F, Stein P, Finkelmann H (2000) Measurement of the pyroelectric response and of the thermal diffusivity of microtomized sections of 'single crystalline' ferroelectric liquid crystalline elastomers. Liq Cryst 27:289
128. Roy SS, Lehmann W, Gebhard E, Tolksdorf C, Zentel R, Kremer F (2002) Inverse piezoelectric and electrostrictive response in freely suspended FLC elastomer film as detected by interferometric measurements. Molec Cryst Liq Cryst 375:253
129. Shenoy DK, Thomsen DL, Srinivasan A, Keller P, Ratna BR (2002) Carbon coated liquid crystal elastomer film for artificial muscle applications. Sens Actuators A 96:184
130. Skupin H, Kremer F, Shilov SV, Stein P, Finkelmann H (1999) Time-resolved FTIR spectroscopy on structure and mobility of single crystal ferroelectric liquid crystalline elastomers. Macromolecules 32:3746
131. Huang CH, Zhang Q, Jakli A (2003) Nematic anisotropic liquid-crystal gels–self-assembled nanocomposites with high electromechanical reponse. Adv Funct Mater 13:525
132. Finkelmann H, Shahinpoor M (2002) Electrically controllable liquid crystal elastomer-graphite composite artificial muscles. Proc SPIE 4695:459
133. Finkelmann H, Wermter H (2000) LC-elastomers as artificial muscles. ACS Abstr 219:189
134. Ahir SV, Tajbakhsh AR, Terentjev EM (2006) Self-assembled shape-memory fibers of triblock liquid-crystal polymers. Adv Funct Mater 16:556
135. Naciri J, Srinivasan A, Joen H, Nikolov N, Keller P, Ratna BR (2003) Nematic elastomer fiber actuators. Macromolecules 36:8499
136. Chambers M, Finkelmann H, Remskar M, Sánchez-Ferrer A, Zalar B, Zumer S (2009) Liquid crystal elastomer-nanoparticle systems for actuation. J Mater Chem 19:1524
137. Spillmann CM, Ratna BR, Naciri J (2007) Anisotropic actuation in electroclinic liquid crystal elastomers. Appl Phys Lett 90:021911
138. Walba DM, Yang H, Shoemaker RK, Keller P, Shao R, Coleman DA, Jones CD, Nakata M, Clark NA (2006) Main-chain chiral smectic polymers showing a large electroclinic effect in the SmA* phase. Chem Mater 18:4576
139. Yu Z, Yuan W, Brochu P, Chen B, Liu Z, Pei Q (2009) Large-strain, rigid-to-rigid deformation of bistable electroactive polymers. Appl Phys Lett 95:192904
140. Yu Z, Niu X, Brochu P, Yuan W, Li H, Chen B, Pei Q (2010) Bistable electroactive polymers (BSEP): large-strain actuation of rigid polymers. Proc SPIE 7642:76420C
141. Pelrine RE, Kornbluh RD, Joseph JP (1998) Electrostriction of polymer dielectrics with compliant electrodes as a means of actuation. Sens Actuators A 64:77
142. Krakovsky I, Romjin T, Posthuma de Boer A (1999) A few remarks on the electrostriction of elastomers. J Appl Phys 85:628

143. Pelrine R, Kornbluh R, Pei Q, Joseph J (2000) High-speed electrically actuated elastomers with strain greater than 100%. Science 287:836
144. Díaz-Calleja R, Sanchis MJ, Riande E (2009) Effect of an electric field on the deformation of incompressible rubbers: bifurcation phenomena. J Electrost 67:158
145. Goulbourne N, Mockensturm E, Frecker M (2005) A nonlinear model for dielectric elastomer membranes. Modeling of a pre-strained circular actuator made of dielectric elastomers. J Appl Mech 72:899
146. Wissler M, Mazza E (2005) Modeling of a pre-strained circular actuator made of dielectric elastomers. Sens Actuators A 120:184
147. Kofod G (2001) Dielectric elastomer actuators. Ph.D. Thesis, The Technical University of Denmark, Sept 2001
148. Wissler M, Mazza E (2005) Modeling and simulation of dielectric elastomer actuators. Smart Mater Struct 14:1396
149. Sommer-Larsen P, Kofod G, Shridhar MH, Benslimane M, Gravesen P (2002) Performance of dielectric elastomer actuators and materials. Proc SPIE 4695:158
150. Yang E, Frecker M, Mockensturm E (2005) Viscoelastic model of dielectric elastomer membranes. Proc SPIE 5759:82
151. Plante JS, Dubowsky S (2007) On the performance mechanisms of dielectric elastomer actuators. Sens Actuators A 137:96
152. Wissler M, Mazza E (2007) Electromechanical coupling in dielectric elastomer actuators. Sens Actuators A 138:384
153. Hwang HW, Kim CJ, Kim SJ, Yang H, Park NC, Park YP (2008) Preisach modeling of dielectric elastomer EAP actuator. Proc SPIE 6927:692726
154. Zhao X, Suo Z (2008) Electrostriction in elastic dielectrics undergoing large deformation. J Appl Phys 104:123530
155. Fox JW, Goulbourne NC (2008) On the dynamic electromechanical loading of dielectric elastomer membranes. J Mech Phys Solids 56:2669
156. Fox JW, Goulbourne NC (2008) Nonlinear dynamic characteristics of dielectric elastomer membranes. Proc SPIE 6927:69271P
157. Kornbluh R, Pelrine R, Joseph J, Heydt R, Pei Q, Chiba S (1999) High-field electrostriction of elastomeric polymer dielectrics for actuation. Proc SPIE 3669:149
158. Zhou J, Hong W, Zhao X, Zhang Z, Suo Z (2008) Propagation of instability in dielectric elastomers. Int J Solids Struct 45:3739
159. Moscardo M, Zhao X, Suo Z, Lapusta Y (2008) On designing dielectric elastomer actuators. J Appl Phys 104:093503
160. Plante JS, Dubowsky S (2006) Large-scale failure modes of dielectric elastomer actuators. Int J Solids Struct 43:7727
161. Zhao X, Hong W, Suo Z (2007) Electromechanical hysteresis and coexistent states in dielectric elastomers. Phys Rev B 76:134113
162. Díaz-Calleja R, Riande E, Sanchis MJ (2008) On electromechanical stability of dielectric elastomers. Appl Phys Lett 93:101902
163. Leng J, Liu L, Liu Y, Yu K, Sun S (2009) Electromechanical stability of dielectric elastomer. Appl Phys Lett 94:211901
164. Pelrine R, Kornbluh R, Joseph J, Chiba S (1997) Electrostriction of polymer films for microactuators. IEEE Tenth Annual International Workshop on MEMS 238
165. Pelrine R, Kornbluh R, Kofod G (2000) High-strain actuator materials based on dielectric elastomers. Adv Mater 12:1223
166. Kornbluh R, Pelrine R, Pei Q, Oh S, Joseph J (2000) Ultrahigh strain response of field-actuated elastomeric polymers. Proc SPIE 3987:51
167. Carpi F, Mazzoldi A, De Rossi D (2003) High-strain dielectric elastomer actuation. Proc SPIE 5051:419
168. Kofod G, Sommer-Larsen P, Kornbluh R, Pelrine R (2003) Actuation response of polyacrylate dielectric elastomers. J Intell Mater Syst Struct 14:787

169. Kofod G, Kornbluh R, Pelrine R, Sommer-Larsen P (2001) Actuation response of polyacrylate dielectric elastomers. Proc SPIE 4329:141
170. Zhao X, Suo Z (2007) Method to analyze electromechanical stability of dielectric elastomers. Appl Phys Lett 91:061921
171. Palakodeti R, Kessler MR (2006) Influence of frequency and prestrain on the mechanical efficiency of dielectric electroactive polymer actuators. Mater Lett 60:3437
172. Choi HR, Jung K, Chuc NH, Jung M, Koo I, Koo J, Lee J, Lee J, Nam J, Cho M, Lee Y (2005) Effects of prestrain on behavior of dielectric elastomer actuator. Proc SPIE 5759:283
173. Lochmatter P, Kovacs G, Wissler M (2007) Characterization of dielectric elastomer actuators based on a visco-hyperelastic film model. Smart Mater Struct 16:477
174. Wissler A, Mazza E (2005) Modeling of a pre-strained circular actuator made of dielectric elastomers. Sens Actuators A 120:184
175. Pei Q, Pelrine R, Stanford S, Kornbluh R, Rosenthal M (2003) Electroelastomer rolls and their application for biomimetic walking robots. Synth Met 135–136:129
176. Goulbourne NC, Mockensturm EM, Frecker MI (2007) Electro-elastomers: large deformation analysis of silicone membranes. Int J Solids Struct 44:2609
177. Goulbourne N, Mockensturm E, Frecker M (2005) A nonlinear model for dielectric elastomer membranes. J Appl Mech 72:899
178. Lochmatter P, Kovacs G (2008) Design and characterization of an active hinge segment based on soft dielectric EAPs. Sens Actuators A 141:577
179. Lochmatter P, Kovacs G, Silvain M (2007) Characterization of dielectric elastomer actuators based on a hyperelastic film model. Sens Actuators A 135:748
180. Mockensturm EM, Goulbourne N (2006) Dynamic response of dielectric elastomers. Int J Non-Linear Mech 41:388
181. Wissler M, Mazza E (2007) Mechanical behavior of an acrylic elastomer used in dielectric elastomer actuators. Sens Actuators A 134:494
182. Wissler M, Mazza E (2007) Electromechanical coupling in dielectric elastomer actuators. Sens Actuators A 138:384
183. Kofod G (2008) The static actuation of dielectric elastomer actuators: how does pre-stretch improve actuation?. J Phys D Appl Phys 41:215405
184. Huang C, Zhang QM, deBotton G, Bhattacharya K (2004) All-organic dielectric-percolative three-component composite materials with high electromechanical response. Appl. Phys Lett 84:4391
185. Gallone G, Carpi F, De Rossi D, Levita G, Marchetti A (2007) Dielectric constant enhancement in a silicone elastomer filled with lead magnesium niobate-lead titanate. Mater Sci Eng C 27:110
186. Zhang Z, Liu L, Fan J, Yu K, Shi L, Leng J (2008) New silicone dielectric elastomers with a high dielectric constant. Proc SPIE 6926:692610
187. Lotz P, Matysek M, Lechner P, Hamann M, Schlaak HF (2008) Dielectric elastomer actuators using improved thin film processing and nanosized particles. Proc SPIE 6927:692723
188. Szabo JP, Hiltz JA, Cameron CG, Underhill RS, Massey J, White B, Leidner J (2003) Elastomeric composites with high dielectric constant for use in Maxwell stress actuators. Proc SPIE 5051:180
189. Carpi F, De Rossi D (2005) Improvement of electromechanical actuating performances of a silicone dielectric elastomer by dispersion of titanium dioxide powder. IEEE Trans Dielectr Electr Insul 12:835
190. Razzaghi-Kashani M, Gharavi N, Javadi S (2008) The effects of organo-clay on the dielectric properties of silicone rubber. Smart Mater Struct 17:065035
191. Zhang XQ, Wissler M, Jaehne B, Broennimann R, Kovacs G (2004) Effects of crosslinking, prestrain, and dielectric filler on the electromechanical response of a new silicone and comparison with acrylic elastomer. Proc SPIE 5385:78

192. Wichiansee W, Sirivat A (2009) Electrorheological properties of poly(dimethylsiloxane) and poly(3,4- ethylenedioxy thiophene)/poly(stylene sulfonic acid)/ethylene glycol blends. Mater Sci Eng C 29:78
193. Carpi F, Gallone G, Galantini F, De Rossi D (2008) Silicone-poly(hexylthiophene) blends as elastoemrs with enhanced electromechanical transduction properties. Adv Funct Mater 18:235
194. Chen LZ, Chen CH, Hu CH, Fan SS (2008) Electrothermal actuation based on carbon nanotube network in silicone elastomer. Appl Phys Lett 92:263104
195. Pei Q, Pelrine R, Rosenthal M, Stanford S, Prahlad H, Kornbluh R (2004) Recent progress on electroelastomer artificial muscles and their application for biomimetic robots. Proc SPIE 5385:41
196. Mathew G, Rhee JM, Nah C, Leo DJ, Nah C (2006) Effects of silicone rubber on properties of dielectric acrylate elastomer actuator. Polym Eng Sci 46:1455
197. Ha SM, Yuan W, Pei Q, Pelrine R, Stanford S (2006) Interpenetrating polymer networks for high-performance electroelastomer artificial muscles. Adv Mater 18:887
198. Ha SM, Yuan W, Pei Q, Pelrine R, Stanford S (2006) New high-performance electroelastomer based on interpenetrating polymer networks. Proc SPIE 6168:616808
199. Ha SM, Yuan W, Pei Q, Pelrine R, Stanford S (2007) Interpenetrating networks of elastomers exhibiting 300% electrically-induced area strain. Smart Mater Struct 16:S280
200. Ha SM, Wissler M, Pelrine R, Stanford S, Kovacs G, Pei Q (2007) Characterization of electroelastomers based on interpenetrating polymer networks. Proc SPIE 6524:652408
201. Ha SM, Park IS, Wissler M, Pelrine R, Stanford S, Kim KJ, Kovacs G, Pei Q (2008) High electromechanical performance of electroelastomers based on interpenetrating polymer networks. Proc SPIE 6927:69272C
202. Shankar R, Ghosh TK, Spontak RJ (2007) Electroactive nanostructured polymers as tunable actuators. Adv Mater 19:2218
203. Shankar R, Ghosh TK, Spontak RJ (2007) Electromechanical response of nanostructured polymer systems with no mechanical pre-strain. Macromol Rapid Commun 28:1142
204. Shankar R, Ghosh TK, Spontak RJ (2009) Mechanical and actuation behavior of electroactive nanostructured polymers. Sens Actuators A 151:46
205. Zhang QM, Su J, Kim CH, Ting R, Capps R (1997) An experimental investigation of electromechanical responses in a polyurethane elastomer. J Appl Phys 81:2770
206. Cameron CG, Underhill RS, Rawji M, Szabo JP (2004) Conductive filler: elastomer composites for Maxwell stress actuator applications. Proc SPIE 5385:51
207. Nam JD, Hwang SD, Choi HR, Lee JH, Kim KJ, Heo S (2005) Electrostrictive polymer nanocomposites exhibiting tunable electrical properties. Smart Mater Struct 14:87
208. Guiffard B, Seveyrat L, Sebald G, Guyomar D (2006) Enhanced electric field-induced strain in non-percolative carbon nanopowder/polyurethane composites. J Phys D Appl Phys 39:3053
209. Petit L, Guiffard B, Seveyrat L, Guyomar D (2008) Actuating abilities of electroactive carbon nanopowder/polyurethane composite films. Sens Actuators A 148:105
210. Zhang QM, Li H, Poh M, Xia F, Cheng ZY, Xu H, Huang C (2002) An all-organic composite actuator material with a high dielectric constant. Nature 419:284
211. Jung K, Lee JH, Cho MS, Koo JC, Nam J, Lee YK, Choi HR (2006) Development of enhanced synthetic rubber for energy efficient polymer actuators. Proc SPIE 6168:61680N
212. Jung K, Lee J, Cho M, Koo JC, Nam J, Lee Y, Choi HR (2007) Development of enhanced synthetic elastomer for energy-efficient polymer actuators. Smart Mater Struct 16:S288
213. Nguyen HC, Doan VT, Park J, Koo JC, Lee Y, Nam J, Choi HR (2009) The effects of additives on the actuating performances of a dielectric elastomer actuator. Smart Mater Struct 18:015006
214. Carpi F, Chiarelli P, Mazzoldi A, De Rossi D (2003) Electromechanical characterization of dielectric elastomer planar actuators: comparative evaluation of different electrode materials and different counterloads. Sens Actuators A 107:85

215. Benslimane M, Gravesen P (2002) Mechanical properties of dielectric elastomer actuators with smart metallic compliant electrodes. Proc SPIE 4695:150
216. Lacour SP, Wagner S, Huang Z, Suo Z (2003) Stretchable gold conductors on elastomeric substrates. Appl Phys Lett 82:2404
217. Lalli JH, Hannah S, Bortner M, Subrahmanyan S, Mecham J, Davis B, Claus RO (2004) Self-assembled nanostructured conducting elastomeric electrodes. Proc SPIE 5385:290
218. Hill AB, Claus RO, Lalli JH, Mecham JB, Davis BA, Goff RM, Subrahmanyan S (2005) Metal rubber electrodes for active polymer devices. Proc SPIE 5759:246
219. Claus RO, Goff RM, Homer M, Hill AB, Lalli JH (2006) Ultralow modulus electrically conducting electrode materials. Proc SPIE 6168:61681O
220. Delille R, Urdaneta M, Hseih K, Smela E (2006) Novel compliant electrodes based on platinum salt reduction. Proc SPIE 6168:61681Q
221. Rosset S, Niklaus M, Dubois P, Shea HR (2008) Mechanical characterization of a dielectric elastomer microactuator with ion-implanted electrodes. Sens Actuators A 144:185
222. Rosset S, Niklaus M, Stojanov V, Felber A, Dubois P, Shea HR (2008) Ion-implanted compliant and patternable electrodes for miniaturized dielectric elastomer actuators. Proc SPIE 6927:69270W
223. Rosset S, Niklaus M, Dubois P, Shea HR (2008) Performance characterization of miniaturized dielectric elastomer actuators fabricated using metal ion implantation. IEEE MEMS 2008, Tucson, AZ, USA, 13–17 Jan 2008, p 503
224. Rosset S, Niklaus M, Dubois P, Shea HR (2009) Metal ion implantation for the fabrication of stretchable electrodes on elastomers. Adv Funct Mater 19:470
225. Yuan W, Lam T, Biggs J, Hu L, Yu Z, Ha SM, Xi D, Senesky MK, Grűner G, Pei Q (2007) New electrode materials for dielectric elastomer actuators. Proc SPIE 6524:65240N
226. Lam T, Tran H, Yuan W, Yu Z, Ha SM, Kaner R, Pei Q (2008) Polyaniline nanofibers as a novel electrode material for fault-tolerant dielectric elastomer actuators. Proc SPIE 6927:69270O
227. Yuan W, Hu L, Ha SM, Lam T, Grűner G, Pei Q (2008) Self-clearable carbon nanotube electrodes for improved performance of dielectric elastomer actuators. Proc SPIE 6927:69270P
228. Yuan W, Hu L, Yu Z, Lam T, Biggs J, Ha SM, Xi D, Chen B, Senesky MK, Grűner G, Pei Q (2008) Fault-tolerant dielectric elastomer actuators using single-walled carbon nanotube electrodes. Adv Mater 20:621
229. Hu L, Yuan W, Brochu P, Gruner G, Pei Q (2009) Highly stretchable, conductive, and transparent nanotube thin films. Appl Phys Lett 94:161108
230. Yuan W, Brochu P, Zhang H, Jan A, Pei Q (2009) Long lifetime dielectric elastomer actuators under continuous high strain actuation. Proc SPIE 7287:72870O
231. Kornbluh R, Pelrine R, Joseph J (1995) Elastomeric dielectric artificial muscle actuators for small robots. Proceedings of the 3rd IASTED international conference on robotics and manufacturing. Cancun, Mexico, (ACTA Press, Calgary, Alberta) 14–16 June 1995 pp 1–6
232. Kornbluh R, Pelrine R, Eckerle J, Joseph J (1998) Electrostrictive polymer artificial muscle actuators. Proceedings of the 1998 IEEE international conference on robotics and automation. Leuven, Belgium, (IEEE Press, Piscataway, NJ) May 1998 pp 2147–2154
233. Carpi F, De Rossi D, Kornbluh R, Pelrine R, Sommer-Larsen P (eds) (2008) Dielectric elastomers as electromechanical transducers: fundamentals, materials, devices, models and applications of an emerging electroactive polymer technology. Elsevier Ltd., Oxford
234. Choi H, Ryew S, Jung K, Jeon J, Kim H, Nam J, Takanishi A, Maeda R, Kaneko K, Tanie K (2002) Biomimetic actuator based on dielectric polymer. Proc SPIE 4695:138
235. Choi HR, Jung KM, Kwak JW, Lee SW, Kim HM, Jeon JW, Nam JD (2003) Multiple degree-of-freedom digital soft actuator for robotic applications. Proc SPIE 5051:262
236. Jung MY, Chuc NH, Kim JW, Koo IM, Jung KM, Lee YK, Nam JD, Choi HR, Koo JC (2006) Fabrication and characterization of linear motion dielectric elastomer actuators. Proc SPIE 6168:616824

237. Goulbourne NCS (2006) Cylindrical dielectric elastomer actuators reinforced with inextensible fibers. Proc SPIE 6168:61680A
238. Lochmatter P, Kovacs G (2007) Concept study on active shells driven by soft dielectric EAP. Proc SPIE 6524:65241O
239. Wang H, Zhu J (2008) Implementation and simulation of a cone dielectric elastomer actuator. Proc SPIE 7266:726607
240. Lenarcic J, Wenger P (eds) (2008) Advances in robot kinematics: analysis and design. Springer Science + Business Media B.V, Berlin, pp 291–299
241. Liu Y, Shi L, Liu L, Zhang Z, Leng J (2008) Inflated dielectric elastomer actuator for eyeball's movements: fabrication, analysis and experiments. Proc SPIE 6927:69271A
242. Prahlad H, Pelrine R, Kornbluh R, von Guggenberg P, Chhokar S, Eckerle J (2005) Programmable surface deformation: thickness-mode electroactive polymer actuators and their applications. Proc SPIE 5759:102
243. Carpi F, De Rossi D (2004) Dielectric elastomer cylindrical actuators: electromechanical modeling and experimental evaluation. Mater Sci Eng C 24:555
244. Kofod G, Paajanen M, Bauer S (2006) New design concept for dielectric elastomer actuators. Proc SPIE 6168:61682J
245. Kofod G, Wirges W, Paajanen M, Bauer S (2007) Energy minimization for self-organized structure formation and actuation. Appl Phys Lett 90:081916
246. O'Brien B, Calius E, Xie S, Anderson I (2008) An experimentally validated model of a dielectric elastomer bending actuator. Proc SPIE 6927:69270T
247. Plante JS, Devita LM, Dubowsky S (2007) A road to practical dielectric elastomer actuators based robotics and mechatronics: discrete actuation. Proc SPIE 6524:652406
248. Bonwit N, Heim J, Rosenthal M, Duncheon C, Beavers A (2006) Design of commercial applications of EPAM technology. Proc SPIE 6168:616805
249. Pei Q, Pelrine R, Stanford S, Kornbluh R, Rosenthal M (2003) Electroelastomer rolls and their application for biomimetic walking robots. Synth Met 135–136:129
250. Zhang R, Lochmatter P, Kunz A, Kovacs G (2006) Spring roll dielectric elastomer actuators for a portable force feedback glove. Proc SPIE 6168:61681T
251. Pei Q, Rosenthal M, Pelrine R, Stanford S, Kornbluh R (2003) Multifunctional electroelastomer roll actuators and their application for biomimetic walking robots. Proc SPIE 5051:281
252. Pei Q, Rosenthal M, Stanford S, Prahlad H, Pelrine R (2004) Multiple-degrees-of-freedom electroelastomer roll actuators. Smart Mater Struct 13:N86
253. Pei Q, Pelrine R, Rosenthal M, Stanford S, Prahlad H, Kornbluh R (2004) Recent progress on electroelastomer artificial muscles and their application for biomimetic robots. Proc SPIE 5385:41
254. Chuc NH, Park J, Thuy DV, Kim HS, Koo J, Lee Y, Nam J, Choi HR (2007) Linear artificial muscle actuator based on synthetic elastomer. Proc SPIE 6524:65240J
255. Carpi F, Migliore A, Serra G, De Rossi D (2005) Helical dielectric elastomer actuators. Smart Mater Struct 14:1210
256. Carpi F, Migliore A, De Rossi D (2005) A new contractile linear actuator made of dielectric elastomers. Proc SPIE 5759:64
257. Carpi F, De Rossi D (2006) Contractile dielectric elastomer actuator with folded shape. Proc SPIE 6168:61680D
258. Carpi F, De Rossi D (2007) Contractile folded dielectric elastomer actuators. Proc SPIE 6524:65240D
259. Carpi F, Salaris D, De Rossi D (2007) Folded dielectric elastomer actuators. Smart Mater Struct 16:S300
260. Schlaak HF, Jungmann M, Matysek M, Lotz P (2005) Novel multilayer electrostatic solid state actuators with elastic dielectric. Proc SPIE 5759:121
261. Matysek M, Lotz P, Flittner K, Schlaak HF (2008) High-precision characterization of dielectric elastomer stack actuators and their material parameters. Proc SPIE 6927:692722

262. Arora S, Ghosh T, Muth J (2007) Dielectric elastomer based prototype fiber actuators. Sens Actuators A 136:321
263. Cameron CG, Szabo JP, Johnstone S, Massey J, Leidner J (2008) Linear actuation in coextruded dielectric elastomer tubes. Sens Actuators A 147:286
264. Kovacs G, Düring L (2009) Contractive tension force stack actuator based on soft dielectric EAP. Proc SPIE 7287:72870A
265. Kovacs G, Ha SM, Michel S, Pelrine R, Pei Q (2008) Study on core free rolled actuator based on soft dielectric EAP. Proc SPIE 6927:69270X
266. Hanson D, White V (2004) Converging the capabilities of EAP artificial muscles and the requirements of bio-inspired robotics. Proc SPIE 5385:29
267. Kovacs G, Lochmatter P, Wissler M (2007) An arm wrestling robot driven by dielectric elastomer actuators. Smart Mater Struct 16:S306
268. Carpi F, De Rossi D (2005) Eyeball pseudo-muscular actuators for an android face. Proc SPIE 5759:16
269. Carpi F, Fantoni G, Guerrini P, De Rossi D (2006) Buckling dielectric elastomer actuators and their use as motors for the eyeballs of an android face. Proc SPIE 6168:61681A
270. Biddis E, Chau T (2008) Dielectric elastomers as actuators for upper limb prosthetics: challenges and opportunities. Med Eng Phys 30:403
271. Dubowsky S, Kesner S, Plante JS, Boston P (2008) Hopping mobility concept for search and rescue robots. Ind Robot Int J 35:238
272. Pelrine R, Sommer-Larsen P, Kornbluh R, Heydt R, Koffod G, Pei Q, Gravesen P (2001) Applications of dielectric elastomer actuators. Proc SPIE 4329:335
273. Choi H, Ryew S, Jung K, Jeon J, Kim H, Nam J, Takanishi A, Maeda R, Kaneko K, Tanie K (2002) Biomimetic actuator based on dielectric polymer. Proc SPIE 4695:138
274. Jung K, Nam H, Lee Y, Choi H (2004) Micro inchworm robot actuated by artificial muscle actuator based on nonprestrained dielectric elastomer. Proc SPIE 5385:357
275. Pelrine R, Kornbluh R, Pei Q, Stanford S, Oh S, Eckerle J (2002) Dielectric elastomer artificial muscle actuators: toward biomimetic motion. Proc SPIE 4695:126
276. Khatib O, Kumar V, Pappas GJ (eds) (2009) Experimental robotics: the 11th international symposium, Springer Tracts in Advanced Robotic, vol 54, pp 25–33
277. Heydt R, Kornbluh R, Eckerle J, Pelrine R (2006) Sound radiation properties of dielectric elastomer electroactive polymer loudspeakers. Proc SPIE 6168:61681M
278. Chiba S, Waki M, Kornbluh R, Pelrine R (2007) Extending applications of dielectric elastomer artificial muscle. Proc SPIE 6524:652424
279. Aschwanden M, Stemmer A (2007) Low voltage, highly tunable diffraction grating based on dielectric elastomer actuators. Proc SPIE 6524:65241N
280. Aschwanden M, Niederer D, Stemmer A (2008) Tunable transmission grating based on dielectric elastomer actuators. Proc SPIE 6927:69271R
281. Kim H, Park J, Chuc NH, Choi HR, Nam JD, Lee Y, Jung HS, Koo JC (2007) Development of dielectric elastomer driven micro-optical zoom lens system. Proc SPIE 6524:65241V
282. Lee S, Jung K, Koo J, Lee S, Choi H, Heon J, Nam J, Choi H (2004) Braille display device using soft actuator. Proc SPIE 5385:368
283. Ren K, Liu S, Lin M, Wang Y, Zhang QM (2007) A compact electroactive polymer actuator suitable for refreshable Braille display. Proc SPIE 6524:65241G
284. Koo IM, Jung K, Koo JC, Nam JD, Lee YK, Choi HR (2008) Development of soft-actuator-based wearable tactile display. IEEE Trans Robotics 24:549
285. Carpi F, Mannini A, De Rossi D (2007) Elastomeric contractile actuators for hand rehabilitation splints. Proc SPIE 6927:692705
286. Carpi F, Khanicheh A, Mavroidis C, De Rossi D (2008) Silicone made contractile dielectric elastomer actuators inside 3-Tesla MRI environment. 2008 IEEE/RSJ international conference on intelligent robots and systems, Nice, France, 22–26 Sept 2008, pp 137–142
287. Carpi F, Khanicheh A, Mavroidis C, De Rossi D (2008) MRI compatibility of silicone-made contractile dielectric elastomer actuators. IEEE/ASME Trans Mechatron 13:370

288. Tadakuma K, DeVita LM, Plante JS, Shaoze Y, Dubowsky S (2008) The experimental study of a precision parallel manipulator with binary actuation: with application to MRI cancer treatment. 2008 IEEE international conference on robotic and automation, Pasadena, CA, USA, 19–23 May 2008, pp 2503–2508
289. Xia F, Tadigadapa S, Zhang QM (2006) Electroactive polymer based microfluidic pump. Sens Actuators 125:346
290. Jhong YY, Huang CM, Hsieh CC, Fu CC (2007) Improvement of viscoelastic effects of dielectric elastomer actuator and its application for valve devices. Proc SPIE 6524:65241Y
291. Bolzmacher C, Biggs J, Srinivasan M (2006) Flexible dielectric elastomer actuators for wearable human-machine interfaces. Proc SPIE 6168:616804
292. Michel S, Dürager C, Zobel M, Fink E (2007) Electroactive polymers as a novel actuator technology for lighter-than-air vehicles. Proc SPIE 6524:65241Q
293. Michel S, Bormann A, Jordi C, Fink E (2008) Feasibility studies for a bionic propulsion system of a blimp based on dielectric elastomers. Proc SPIE 6927:69270S
294. Yang WP, Chen LW (2008) The tunable acoustic band gaps of two-dimensional phononic crystals with a dielectric elastomer cylindrical actutor. Smart Mater Struct 17:015011
295. Yang WP, Wu LY, Chen LW (2008) Refractive and focusing behaviours of tunable sonic crystals with dielectric elastomer cylindrical actuators. J Phys D Appl Phys 41:135408
296. Wu LY, Wu ML, Chen LW (2009) The narrow pass band filter of tunable 1D phononic crystals with a dielectric elastomer layer. Smart Mater Struct 18:015011
297. Beck M, Fiolka R, Stemmer A (2009) Variable phase retarder made of a dielectric elastomer actuator. Opt Lett 34:803
298. Beck M, Aschwanden M, Stemmer A (2008) Sub-100-nanometer resolution in total internal reflection fluorescence microscopy. J Microsc 232:99
299. Pelrine R, Kornbluh R, Eckerle J, Jeuck P, Oh S, Pei Q, Stanford S (2001) Dielectric elastomers: generator mode fundamentals and applications. Proc SPIE 4329:148
300. US 6,768,246 B2 (2004) Biologically powered electroactive polymer generators. SRI international, Pelrine RE, Kornbluh RD, Eckerle JS, Stanford SE, Oh S, Garcia PE
301. WO 2007/130252 A2 (2007) Wave powered generation using electroactive polymers. SRI international, Kornbluh RD, Pelrine RE, Prahlad H, Chiba S, Eckerle J, Chavez B, Stanford SE, Low T
302. WO 2007/130253 A2 (2007) Wave powered generation. SRI international, Kornbluh RD, Pelrine RE, Prahlad H, Chiba S, Eckerle J, Chavez B, Stanford SE, Low T
303. Chiba S, Waki M, Kornbluh R, Pelrine R (2008) Innovative power generators for energy harvesting using electroactive polymer artificial muscles. Proc SPIE 6927:692715
304. Jean-Mistral C, Basrour S, Chaillout JJ (2008) Dielectric polymer: scavenging energy from human motion. Proc SPIE 6927:692716
305. Waki M, Chiba S, Kornbluh R, Pelrine R, Kunihiko U (2008) Electric power from artificial muscles. OCEANS 2008—MTS/IEEE Kobe Techno-Ocean, 8–11 Apr 2008, pp 1–3
306. Koh SJA, Zhao X, Suo Z (2009) Maximal energy that can be converted by a dielectric elastomer generator. Appl Phys Lett 94:262902
307. Jean-Mistral C, Basrour S, Chaillout JJ, Bonvilain A (2007) A complete study of electroactive polymers for energy scavenging: modeling and experiments. DTIP 2007, Stresa, Italy, 25–27 Apr 2007
308. Ihlefeld CM, Qu Z (2008) A dielectric electroactive polymer generator-actuator model: modeling, identification, and dynamic simulation. Proc SPIE 6927:69270R
309. Rosenthal M, Bonwit N, Duncheon C, Heim J (2007) Applications of dielectric elastomer EPAM sensors. Proc SPIE 6524:65241F
310. O'Brien B, Thode J, Anderson I (2007) Integrated extension sensor based on resistance and voltage measurement for a dielectric elastomer. Proc SPIE 6524:652415
311. Keplinger C, Kaltenbrunner M, Arnold N, Bauer S (2008) Capacitive extensometry for transient strain analysis of dielectric elastomer actuators. Appl Phys Lett 92:192903
312. Jung K, Kim KJ, Choi HR (2008) A self-sensing dielectric elastomer actuator. Sens Actuators A 143:343

313. Jung K, Kim KJ, Choi HR (2008) Self-sensing of dielectric elastomer actuator. Proc SPIE 6927:69271S
314. Chuc NH, Thuy DV, Park J, Kim D, Koo J, Lee Y, Nam JD, Choi HR (2008) A dielectric elastomer actuator with self-sensing capability. Proc SPIE 6927:69270V

Chapter 2
Modeling of IPMC Guide Wire Stirrer in Endovascular Surgery

Yousef Bahramzadeh and Mohsen Shahinpoor

Abstract Reported are modeling, design, development, and laboratory testing of an electrically bendable soft distal tip of a guide wire or micro-catheter for use during endovascular surgical operations or diagnostics. The catheter distal tip is equipped with an electrically controllable, relatively floppy stirrer. The distal end is equipped with an ionic polymeric metal composite (IPMC) artificial muscle. The artificial muscle is connected to an external voltage and controllable by a feed forward or feedback controller and may be manipulated through a surgical robotic system similar to those used in master–slave robots. By varying the voltage or current applied to the polymeric artificial muscle bender, it can be made to bend/stir in one direction or another. By pointing the distal tip in a desired direction, the catheter or guide wire can be advanced to a specific location such as a cerebral aneurysm in the brain vasculature through body lumens. The artificial muscle distal tip bender is multifunctional or smart because it can serve both as an endovascular actuator and a sensor for hemodynamic flow and force measurement.

Keywords Artificial muscle · IPMC · Actuator · Sensor · Endovascular surgery · Catheter · MIS · SMA · EAP

Y. Bahramzadeh · M. Shahinpoor (✉)
Department of Mechanical Engineering, Biomedical Engineering Laboratory,
University of Maine, Orono, ME 04469, USA
e-mail: mohsen.shahinpoor@maine.edu

L. Rasmussen (ed.), *Electroactivity in Polymeric Materials*,
DOI: 10.1007/978-1-4614-0878-9_2,
© Springer Science+Business Media New York 2012

2.1 Introduction

There is an increasing trend toward replacing the conventional methods of open surgeries with minimally invasive procedures called robotic surgery. These novel techniques give the surgeon new capabilities for manipulating surgical instruments, enabling the surgeon to carry out more complex surgical procedures with enhanced precision. In addition, the amount of pain, recovery period, and hospitalization time is significantly reduced for the patient in Minimally Invasive Methods (MIS) of surgery. Advances in developing new MIS methods depend heavily on developing new materials that are biocompatible and able to satisfy new biomedical requirements [1]. There are numerous surgical applications that require using flexible soft actuators with low actuation power. Also, the possibility of actuators to be miniaturized for operations in small cavities, such as inside the arteries and veins, can extend their biomedical applications and even result in developing new types of MIS methods. An example of these biocompatible actuators is Shape Memory Alloy (SMA) wires, such as Nitinol™, that undergo large strains by controlling the temperature. Another example is Electro Active Polymers (EAPs) that act as low density, resilient artificial muscles where the shape can be controlled by applying an electrical field. Both of these materials have found several commercial biomedical applications [2–5].

Ionic Polymeric Metallic Composites (IPMCs) are a class of EAPs that exhibit characteristics of both actuators and sensors, Shahinpoor et al. [6–11]. The flexibility of an IPMC makes it possible to be applied both in small and large deflection applications. Successive photographs of an IPMC strip are shown in Fig. 2.1 that demonstrates very large deformation (up to 8 cm) in the presence of low voltage. The sample is 10 mm wide, 80 mm long, and 0.34 mm thick. The time interval is 1 s and the actuation voltage is 4 V DC.

High flexibility, low drive voltage, and large bending deflection are definite advantages of IPMCs over other rigid piezoelectric ceramic materials. These characteristics make IPMC actuators and sensors very popular in various biomedical applications.

2.2 Modeling of Actuation and Sensing Mechanisms

Sensing and actuation behaviors observed in IPMC artificial muscles are due to the ion migration throughout the membrane thickness. In Fig. 2.2, the actuation and sensing mechanisms of IPMC artificial muscles are shown. Nano-actuation is imposed by the electric field, which moves the cations towards the cathode, thus causing bending. In a reverse process, nano-sensing is observed when mechanical deformation causes cations to redistribute, thus causing a streaming potential and an electric field.

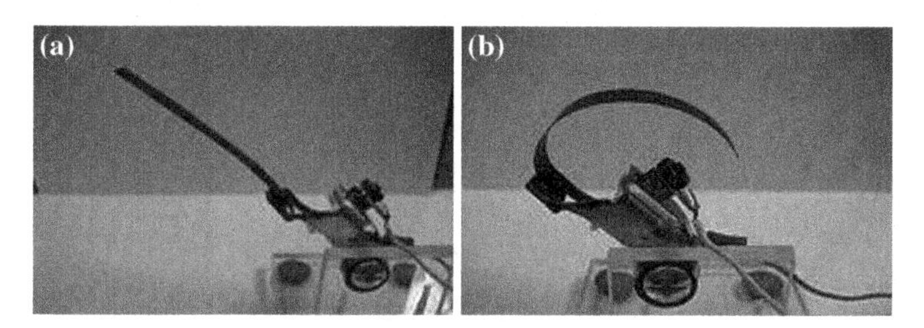

Fig. 2.1 Successive photographs of an IPMC strip before actuation (**a**) and (**b**) after actuation [9]. Smart Materials and Structures 2004, reprinted with permission

Typically, the tip of the perfluorinated ionic polymer strip bends toward the anode (in case of cation exchange membranes) under the influence of an electric potential. Also, the appearance of water on the surface of the expansion side and the disappearance of water on the contraction side occurs near the electrodes.

Let us now summarize the underlying principle of the IPMC's actuation and sensing capabilities, which can be described by the standard Onsager formula using linear irreversible thermodynamics. When static conditions are imposed, a simple description of mechanoelectric effect is possible based upon two forms of transport: ion transport (with a current density, $\underline{J}$, normal to the material) and solvent transport (with a flux, $\underline{Q}$, that we can assume is water flux).

The conjugate forces include the electric field, $\underline{E}$, and the pressure gradient, $\underline{\nabla}_p$. The resulting equation has the concise form of (2.1):

$$\underline{J}(x,y,z,t) = \sigma\underline{E}(x,y,z,t) - L_{12}\underline{\nabla}_p(x,y,z,t) \tag{2.1}$$

$$\underline{Q}(x,y,z,t,) = L_{21}\underline{E}(x,y,z,t) - K\underline{\nabla}_p(x,y,z,t) \tag{2.2}$$

where σ and K are the material electric conductance and the Darcy permeability, respectively. A cross coefficient is usually $L = L_{12} = L_{21}$. The simplicity of the above equations provides a compact view of the underlying principles of actuation, transduction, and sensing of the IPMCs.

Figure 2.2 depicts the redistribution of ions and polar fluids (water) in IPMCs under an imposed electric field. When the *direct* effect (actuation mode) is investigated; that is, upon application of a voltage, a bending deformation is observed. One notes that ideally $\underline{Q} = 0$ after static deformation sets in. Thus:

$$\underline{\nabla}_p(x,y,z,t) = \frac{L(\underline{E})}{K}\underline{E}(x,y,z,t) \tag{2.3}$$

Note that in this case the Onsager coefficient L is indeed a function of the imposed electric field $\underline{E}$. The pressure gradient $\underline{\nabla}_p(x, y, z, t)$ will, in turn, induce a curvature $\underline{\kappa}$ proportional to $\underline{\nabla}_p(x, y, z, t)$. The relationship between the curvature $\underline{\kappa}$ and pressure gradient $\underline{\nabla}_p(x, y, z, t)$ are well known. Recall that $\underline{\kappa} = \mathbf{M}(\underline{E})/\mathrm{YI}$,

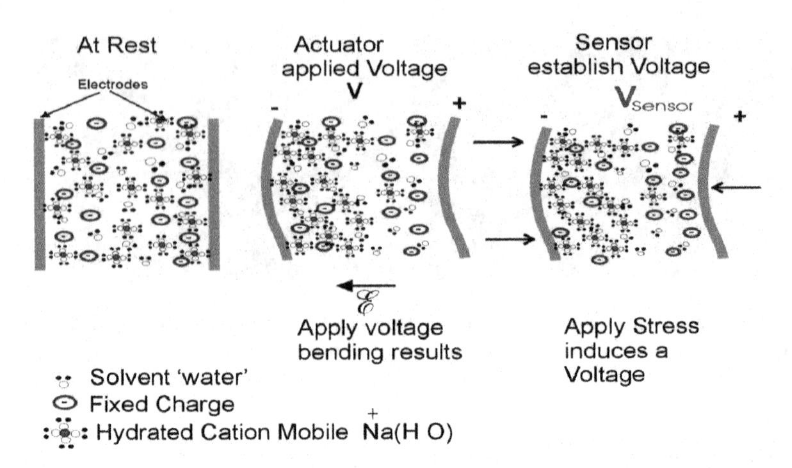

Fig. 2.2 Actuation and sensing mechanism of IPMC actuators and sensors

where **M**($\underline{E}$) is the local induced bending moment and is a function of the imposed electric field $\underline{E}$; Y is the Young's modulus (elastic stiffness) of the strip, which is a function of the hydration of the IPMC; and I is the moment of inertia of the strip. Note that locally **M**($\underline{E}$) is related to the pressure gradient such that in a simplified scalar format:

$$\underline{\nabla}_p(x, y, z, t)\,(\mathbf{M}/\mathrm{I}) = \mathrm{Y}\underline{\kappa}., \tag{2.4}$$

where $\underline{\nabla}_p(x, y, z, t)$ is the pressure gradient or the difference between the tensile and the compressive stresses in the uppermost remote surfaces of the IPMC strip. Now from Eq. (2.4) it is clear that the vector form of curvature $\underline{\kappa}_E$ is related to the imposed electric field $\underline{\kappa}$ by:

$$\underline{\kappa}_E = (L(\underline{E})/\mathrm{KY})\underline{E} \tag{2.5}$$

Based on this simplified model the tip bending deflection $\underline{\delta}_{\mathrm{max}}$ of an IPMC strip of length l_g should be almost linearly related to the imposed electric field due to the fact that:

$$\underline{\kappa}_E \cong [2\underline{\delta}_{\mathrm{max}}/(l_g^2 + \underline{\delta}_{\mathrm{max}}{}^2)] \cong 2\underline{\delta}_{\mathrm{max}}/l_g^2 \cong (L(\underline{E})/\mathrm{KY})\underline{E} \tag{2.6}$$

Here, we have used a low frequency electric field in order to minimize the effect of loose water back diffusion under a step voltage or a DC electric field. Other parameters have been experimentally measured to be $\mathrm{K} \sim 10^{-18}$ m^2/CP, $\sigma \sim$ 1A/mV or S/m. Figure 2.3 depicts a more detailed set of data pertaining to the Onsager coefficient L as a function of electric field $\underline{E}$.

The experimental deformation characteristics are clearly consistent with the above predictions obtained by the above linear irreversible thermodynamics formulation, which are also consistent with Eqs. 2.5 and 2.6 in the steady state conditions, and have been used to estimate the value of the Onsager coefficient L to be of the order of 10^{-8} m^2/V-s (Fig. 2.3).

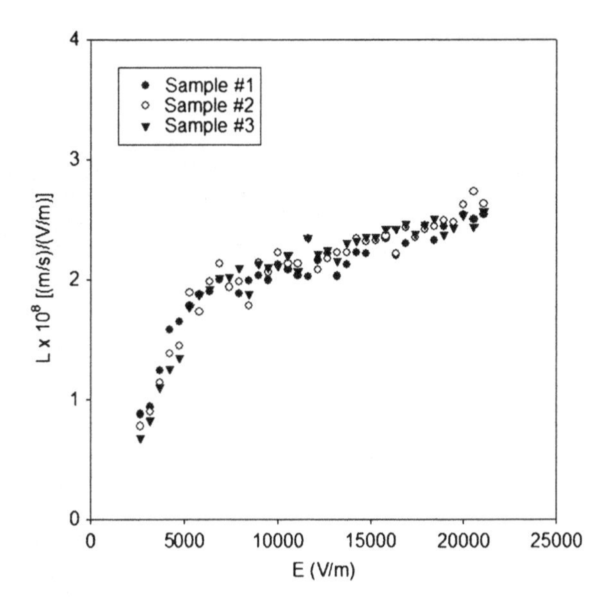

Fig. 2.3 Experimental determination of the Onsager Coefficient *L* using three different samples [9]. Smart Materials and Structures 2004, reprinted with permission

2.3 Endovascular Distal Tip Stirrer

2.3.1 Actuator as an Endovascular Stirrer

Catheter insertion is a minimally invasive surgical process in which a thin wire is inserted into the body through a small incision in the groin or the arm to reach a femoral or brachial artery. The distal tip of a wire is stirred through blood vessels to reach the desired location. Currently, various surgical operations apply catheter insertion techniques for different types of surgeries. In angioplasty for instance, an inflatable balloon at the distal tip of a catheter is used for widening a narrowed or arterial blockage, as in atherosclerosis, stenosis, or atheroma. The process can also be used for placement of a stent, permanently acting like a scaffold for the artery to support the damaged artery walls. Finally, the balloon is deflated and removed from the location. However, guidance of a catheter inside the complex channels of blood vessels or vasculature is a complicated task that is currently performed manually in a trial and error process. The correct insertion depends on the surgeon's skill to a great extent. Using vision feedback from X-ray imaging or fluoroscopy, the surgeon inserts and stirs the distal tip of wire repeatedly to enter different branches in the vasculature. This process should be done very meticulously and carefully in order to prevent further damage to the blood vessels. Very soft stirrers for catheters have been introduced to overcome the aforementioned shortcomings of conventional catheters. Enabling a catheter with one or two controllable degrees of freedom for bending the distal tip will give surgeons more dexterity in maneuvering the wire inside the body. Jayender, Patel, and Nikumb

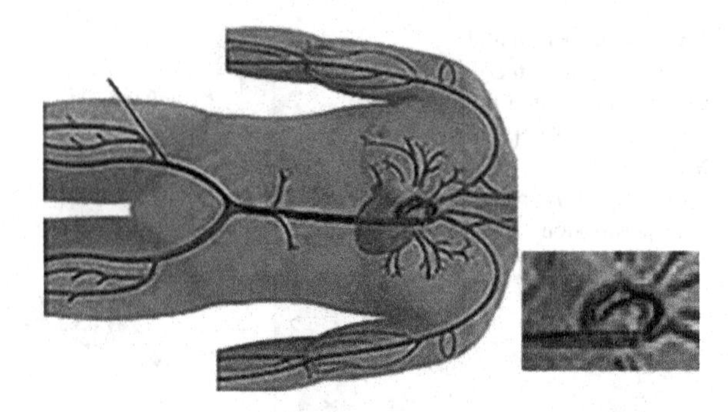

Fig. 2.4 Schematic of the IPMC micro-catheter stirrer

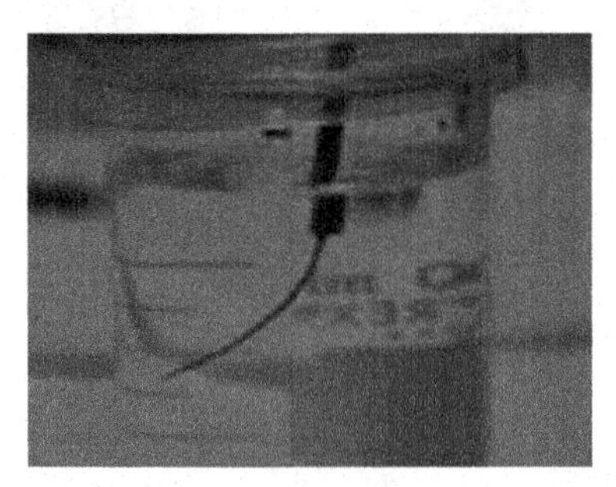

Fig. 2.5 IPMC stirrer attached to the tip of a catheter and actuated in a saline fluidic environment resembling blood

used tendon driven actuation to control the shape of a catheter [3]. Recently, Camarillo, Milne, and Salisbury have proposed using SMA actuators for the bending distal tip [4], while Jayender, Patel, and Nikumb used three SMA wires for 3D orienting of the distal tip in space [5]. Even though SMA actuators provide high stress and strain, high non-linearity, and hysteresis behavior, during the Martensite-Austenite phase transformation causes difficulty in controlling SMA actuators. Since the length of SMA actuators is controlled by Joule heating and cooling processes to induce phase transformation, the dynamic response of these actuators is relatively slow.

In our work, the tip of a catheter is equipped with an IPMC artificial muscle as a stirrer. A schematic of an attached IPMC stirrer to the tip of a catheter is depicted in Fig. 2.4. Bi-directional bending of the catheter, along with the manual twisting motion of the wire, enables 3D orientation control of the active catheter. Successful actuation is achieved in a fluidic environment as shown in Fig. 2.5. The tip of the actuator is easily bent about 90°, which is sufficient to maneuver through the endovascular branches.

The actuation properties of IPMC are presented in. [6–10] By applying a voltage of 0.2–3 V on an IPMC film, bending towards the anode occurred. An increase in voltage level (up to 6 or 7 V) causes larger bending displacement along with nonlinear saturation in displacement. IPMCs also work very well in water or blood environments.

Applying harmonic voltage causes the film to undergo harmonic displacement. The normalized deflection variation with respect to electrical field change is depicted in Fig. 2.6. The displacement amplitude is dependent on both voltage

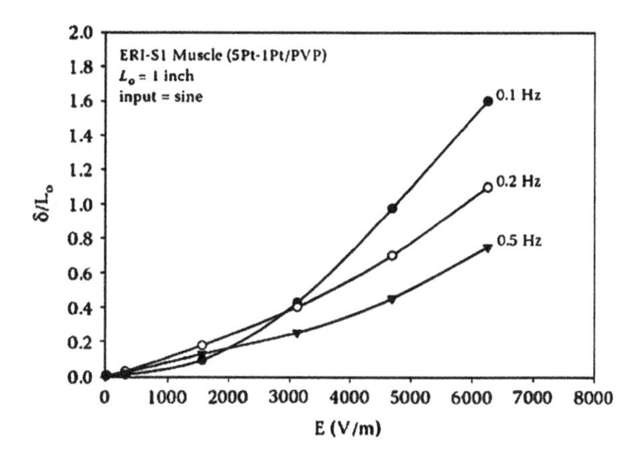

Fig. 2.6 Displacement characterization of an 1in IPMC sheet vs. electric field and frequency [9]. Smart Materials and Structures 2004, reprinted with permission

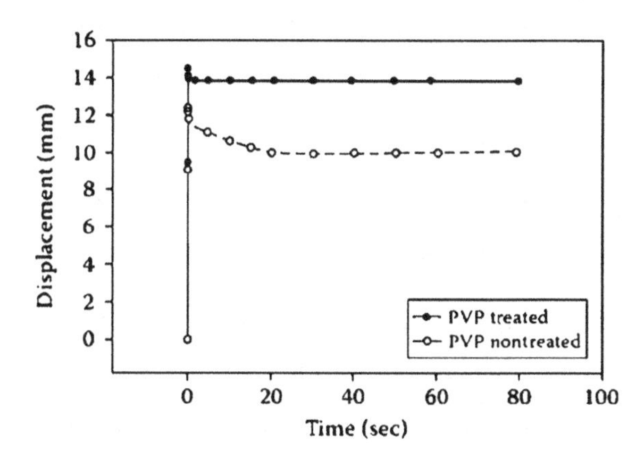

Fig. 2.7 Step voltage response displacement characteristics of IPMNC hydrated samples [9]. Smart Materials and Structures 2004, reprinted with permission

magnitude and frequency. It can be seen that lower frequencies (down to 0.1 or 0.01 Hz) lead to higher displacement.

2.3.2 Modeling of Stirrer for Control Purposes

Different modeling approaches have been introduced, Shahinpoor et al. [6–11]. Fig. 2.7 shows the step response of a 20 × 5 × 0.2 mm IPMC cantilever strip which is actuated by a 1.5 input volt. Two types of IPMCs, PVP (polyvinyl pyrolidone) treated and PVP non-treated, are compared with each other in terms of maximum deflection and settling time. The steady state deflection achieved for the PVP treated sample is about 40% higher than the non-treated sample, and the settling time is reduced significantly, showing fast actuation properties.

Modeling the actuation response of IPMC for robust control of IPMC actuators is described in references 12 and 13. [12–13] Step response of actuators can be used for identification of actuator dynamics to derive a second order transfer

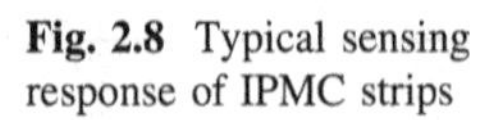

Fig. 2.8 Typical sensing response of IPMC strips

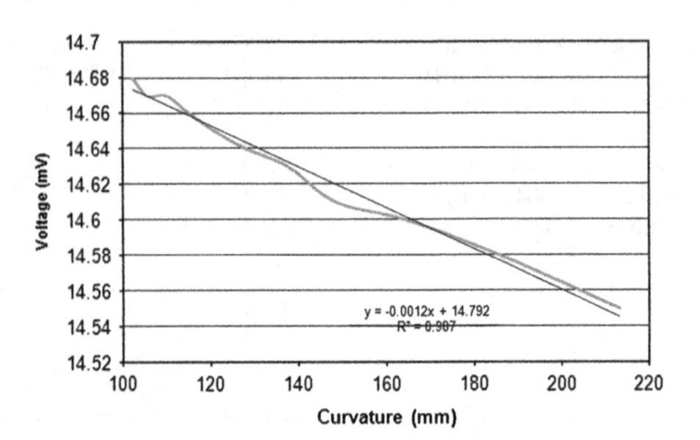

function. Then, a feed forward control method can be applied to control the curvature using a derived transfer function. For higher precision, a feedback control method may be applied, at the cost of adding a vision system, for monitoring the shape of the actuator.

2.3.3 IPMC Sensor for Deflection and Contact Sensing of Endovascular Stirrers

In order for receiving feedback from the deflection of curvature of the IPMC stirrer, one potential possibility is to utilize the sensing property of IPMCs. Figure 2.8 depicts a typical sensing output of IPMC strips. Note that the output voltage of an IPMC sheet is a function of curvature and its rate of change. While an output voltage of 1–2 mV can be derived by dynamic sensing from a sample of size 8 × 20 × 0.2 mm, the achievable voltage is smaller in quasi-static sensing cases using the same sample. Therefore the sensitivity of the sensor, which is enhanced by increasing the frequency and velocity of actuation, should be considered for calibration of an IPMC sensor for deflection measurement of a stirrer. The deflection sensing can be accomplished by attaching another IPMC strip parallel to the IPMC actuator in order to monitor its motion. The size of the sensor should be chosen to minimize the effect on the actuator.

Another possibility is switching between the sensing and actuating functions of the IPMC. The same pair of electrodes can be used along with a switching control to switch from an actuation mode to a sensing mode.

2.4 Conclusion

Modeling and development of an IPMC based distal tip guide wire stirrer were presented. IPMCs can be cut arbitrarily smaller or larger for applications in microelectromechanical systems (MEMS), nano-electromechanical systems (NEMS),

BioMEMS, BioNEMS, or industrial applications requiring soft sensors, actuators, and micro or nano-scale robotic applications.

An IPMC bender was used to equip the distal tip of an endoscopic guide wire or micro-catheter for endovascular navigation to reach the targeted regions or lumens inside the blood vessels and through the tortuously curved paths of blood vasculature. An IPMC strip is a powerful candidate for endovascular catheter navigation because of the following: the high flexibility of IPMC materials; low driving voltage between 0.2 and 4 V; possibility of miniaturization to very small endovascular sizes; successful activation in fluidic media; and almost linear voltage-displacement actuation/sensing behavior. In addition, the sensing property of IPMC artificial muscles can be exploited to equip the distal tip with a curvature sensor as well as a contact sensor to perform the catheter insertion in a less invasive method.

References

1. Mack MJ (2001) Minimally invasive and robotic surgery. JAMA 285(5):568–572. doi:10.1001/jama/285.5.568
2. Fischer H, Vogel B, Pfleging W, Besser H (1999) Flexible distal tip made of nitinol (NiTi) for a steerable endoscopic camera system. Mater Sci Eng A 273–275:780–783
3. Jayender J, Patel RV, Nikumb S (2006) Robot-assisted catheter insertion using hybrid impedance control. In: 2006 IEEE international conference on robotics and automation (ICRA) pp 607–612. doi:10.1109/ROBOT.2006.1641777
4. Camarillo D, Milne C, Salisbury K (2008) Mechanics modeling of tendon driven continuum manipulators. IEEE Trans Robot 24(6):1262–1273. doi:10.1109/TRO.2008.2002311
5. Jayender J, Patel RV, Nikumb S (2009) Robot-assisted active catheter insertion: algorithms and experiments. Int J Robot Res 28(9):1101–1117. doi:10.1177/0278364909103785
6. Shahinpoor M, Kim KJ, Mojarrad M (2007) Artificial muscles, applications of advanced polymeric nanocomposites. CRC Press, Taylor and Francis Publishers, London
7. Shahinpoor M, Kim KJ (2001) Ionic polymer-metal composites I. Fundamentals (review paper). Smart Mater Struct Int J 10(4):819–833. doi:10.1088/0964-1726/10/4/327
8. Kim KJ, Shahinpoor M (2003) Ionic polymer-metal composites II. Manufacturing techniques (review paper). Smart Mater Struct Int J 12(1):65–79. doi:10.1088/0964-1726/12/1/308
9. Shahinpoor M, Kim KJ (2004) Ionic polymer-metal composites III. modeling and simulation as biomimetic sensors, actuators, transducers and artificial muscles (review paper). Smart Mater Struct Int J 13(6):1362–1388. doi:10.1088/0964-1726/13/6/009
10. Shahinpoor M, Kim KJ, (2005) Ionic polymer-metal composites IV. industrial and medical applications (review paper). Smart Mater Struct Int J 14(1):197–214. doi:10.1088/0964-1726/14/1/020
11. De Gennes PG, Okumura K, Shahinpoor M, Kim KJ (2000) Mechanoelectric effects in Ionic Gels. Europhys Lett 50:513–518. doi:10.1209/epl/i2000-00299-3
12. Mallavarapu K, Newbury K, Leo D (2001) Feedback control of the bending response of ionic polymer–metal composite actuators. SPIE Proc 4329:301–310. doi:10.1117/12.432660
13. Lee MJ, Jung SH, Mun MS, Lee S, Moon I (2006) Control of IPMC-based artificial muscle for myoelectric hand prosthesis. Proceedings 1st IEEE/RAS-EMBS international conference on biomedical robotics and biomechatronics, pp 1172–1177. doi:10.1109/BIOROB.2006.1639251

Chapter 3
From Boots to Buoys: Promises and Challenges of Dielectric Elastomer Energy Harvesting

Roy D. Kornbluh, Ron Pelrine, Harsha Prahlad, Annjoe Wong-Foy, Brian McCoy, Susan Kim, Joseph Eckerle and Tom Low

Abstract Dielectric elastomers offer the promise of energy harvesting with few moving parts. Power can be produced simply by stretching and contracting a relatively low-cost rubbery material. This simplicity, combined with demonstrated high energy density and high efficiency, suggests that dielectric elastomers are promising for a wide range of energy harvesting applications. Indeed, dielectric elastomers have been demonstrated to harvest energy from human walking, ocean waves, flowing water, blowing wind, and pushing buttons. While the technology is promising, there are challenges that must be addressed if dielectric elastomers are to be a successful and economically viable energy harvesting technology. These challenges include developing materials and packaging that sustains long lifetime over a range of environmental conditions, design of the devices that stretch the elastomer material, as well as system issues such as practical and efficient energy harvesting circuits. Progress has been made in many of these areas. We have demonstrated energy harvesting transducers that have operated over 5 million cycles. We have also shown the ability of dielectric elastomer material to survive for months underwater while undergoing voltage cycling. We have shown circuits capable of 78% energy harvesting efficiency. While the possibility of long lifetime has been demonstrated at the watt level, reliably scaling up to the power levels required for providing renewable energy to the power grid or for local use will likely require further development from the material through to the systems level.

RD Kornbluh, R Pelrine, H Prahlad, A Wong-Foy, B McCoy, S Kim, J Eckerle, T Low, "From Boots to Buoys: Promises and Challenges of Dielectric Elastomer Energy Harvesting," SPIE Proc 7976: 48–66, Bellingham, WA, 2011 [doi: 10.1117/12.882367], reprinted with permission.

R. D. Kornbluh (✉) · R. Pelrine · H. Prahlad · A. Wong-Foy ·
B. McCoy · S. Kim · J. Eckerle · T. Low
SRI International, 333 Ravenswood Avenue, Menlo Park, CA 94025, USA
e-mail: roy.kornbluh@sri.com

L. Rasmussen (ed.), *Electroactivity in Polymeric Materials*,
DOI: 10.1007/978-1-4614-0878-9_3,
© Springer Science+Business Media New York 2012

Keywords Dielectric elastomer • Electroactive polymer • Transducer • Energy harvesting • Lifetime • Wave powerwave power

3.1 Introduction

Dielectric elastomer transducers, a type of electroactive polymer (EAP), are comprised of deformable polymer films that respond to an electric field applied across their thickness. Dielectric elastomers offer unique properties as an electromechanical transducer technology compared with more conventional transducer technologies such as those based on piezoelectrics or electromagnetics. For example, when acting as an actuator, dielectric transducers have been shown to be capable of large strains (in some cases greater than 100%) [1] with the relatively fast response and high efficiency associated with electric field activated materials [2]. A number of materials can be used for the component materials of dielectric elastomers including those based on relatively inexpensive commercially available materials. The elastomers can be quite soft, suggesting their potential for a variety of applications that may involve human interaction or unusual load-matching requirements. The simple structure, wide availability, and unique properties of dielectric elastomers have allowed researchers to explore their use in a wide variety of actuator applications. Brochu and Pei [3] and Carpi [4] include surveys of state of the art of dielectric elastomers. Although first reported in 2001 [5], the use of dielectric elastomers as generators has been less widespread. Only in the past two years has research in their use as generators seemed to increase dramatically as evidenced by the increase in publications on this topic (many of which are cited herein). This paper seeks to examine the promises and challenges associated with dielectric elastomer energy harvesting. Much of the information in this paper comes from the authors' own experiences developing dielectric elastomer energy harvesting systems for applications including power generating boots and ocean wave power harvesting buoys.

This paper is organized as follows: first, we present a background on the use of the dielectric technology for power generation so that its characteristics and unique properties may be better understood. This background will include a brief comparison with competing energy harvesting technologies of piezoelectrics and electromagnetics. The background will also review the state of the art in our physical understanding of the use of dielectric elastomers in energy harvesting and point out deficiencies that still remain in our understanding. In addition to focusing on the dielectric elastomer technology itself, we will show how the technology must fit in the context of an overall power generation system. These system-related issues include the interface to the mechanical power source, design and material selection of the dielectric elastomer device itself, power harvesting electronics, and economic issues.

Next, we will introduce specific examples of energy harvesting systems that have been built or contemplated by the authors or collaborators. In addition to highlighting the potential of dielectric elastomer power generation, this section will also provide specific examples of the systems-related issues and serve as a springboard for a discussion of these system issues such as lifetime.

The next section focuses in more detail on lifetime issues and their implications for practical adoption of the technology. We highlight examples of long lifetime dielectric elastomer generator elements and discuss the issues involved in achieving such lifetimes and the implications for various generator applications.

We conclude with a discussion that summarizes the challenges that remain and how they can be addressed. It is a thesis of this paper that dielectric elastomer energy harvesting does indeed offer great potential if such challenges can be overcome.

3.2 Background on Dielectric Elastomer Power Generation

3.2.1 Principles of Operation

The basic operational element of a dielectric elastomer generator, shown in Fig. 3.1, is a film of an elastically deformable insulating polymer that is coated on each side by a compliant electrode. In generator mode, dielectric elastomers convert the mechanical work of stretching the polymer film into electrical energy. In order to achieve this conversion, it is necessary to add electrical charge to the surface of the polymer film while it is in a stretched state and allow the elastic forces on the film to relax the film to a state of lower stretch. When the film relaxes it shrinks in area and increases in thickness. If most of the charge on the film is conserved, then both geometric effects tend to increase the electrical energy on the film since like charges on each electrode are forced together while unlike charges on the opposite electrodes are pulled apart. This increase in energy may be greater than that required to initially place the charge on the film, and in some cases can be many times greater.

Note that electrical energy can only be produced if some charge is initially placed on the film. Thus, as shown in Fig. 3.2, a dielectric elastomer generator must include an energy harvesting circuit that controls the adding and removal of the charge. The timing of the application and removal of the charge relative to the stretching and contraction as well as the amount of charge added and removed is critical in determining the amount of energy that can be extracted, as will be discussed below.

The maximum amount of energy that can be converted using a given amount of film depends on the material properties. Several different material properties come into play including the maximum strain that can be imposed before mechanical failure, the maximum electric field that can be supported before electrical breakdown, and the need to maintain elastic restoring forces.

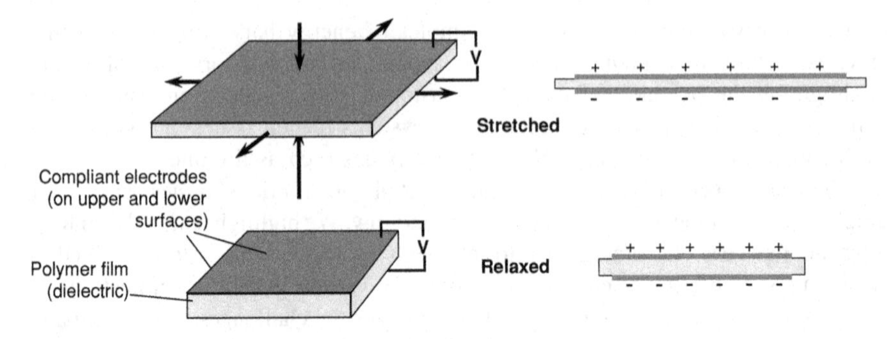

Fig. 3.1 Basic operational element of a dielectric elastomer generator

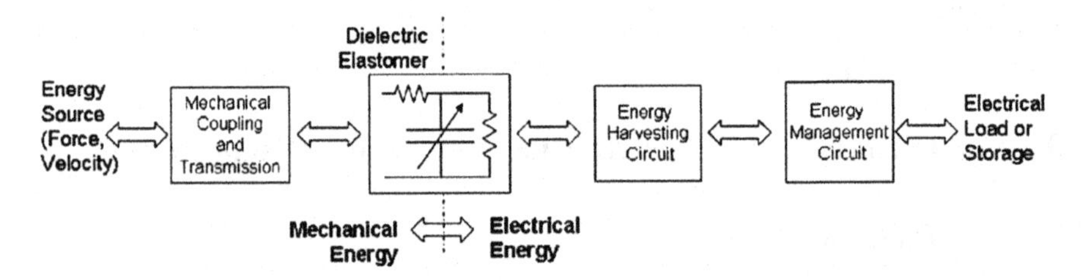

Fig. 3.2 Components of a dielectric elastomer generator system

Koh et al. [6] have rigorously modeled the electromechanics of this interaction for the simplified case of uniform biaxial stretching of an incompressible polymer film including many important effects such as the nonlinear stiffness behavior of the polymer film and the variation in breakdown field with the state of strain. With regard to the latter effect, Pelrine et al. [5] showed the dramatic effect of prestrain on the performance of dielectric elastomers (specifically silicones and acrylics) as actuators. We would expect the same breakdown enhancement effects to be involved with regard to power generation. There are many additional effects that may be important, such as electrical and mechanical loss mechanisms, interaction with the environment or circuits, frequency, and temperature-dependent effects on material parameters. The analysis by Koh provides the state equations

$$F/LH + \varepsilon (V/H)^2 \lambda^3 = s + \varepsilon E^2 \lambda^3 = Y(\lambda) \tag{3.1}$$

$$Q = (\varepsilon L^2 \lambda^{44}/H)V = CV = (\varepsilon A_0 \gamma^4)E, \tag{3.2}$$

where for a polymer film with a constant permittivity of ε, a length and width of L, and an initial (unstretched) thickness H, the state variables are F, λ, V, and Q—the biaxial force on the film, the film stretch ratio in each direction (=1 + strain), the voltage across the film, and the charge on the film, respectively. The nominal stress s relates to the nonlinear elastic behavior of the film $Y(\lambda)$ as $s = Y(\lambda)$. Added to the second equations are expressions that include the more common parameters of the electric field E and the capacitance C as well as the area stretch ratio γ

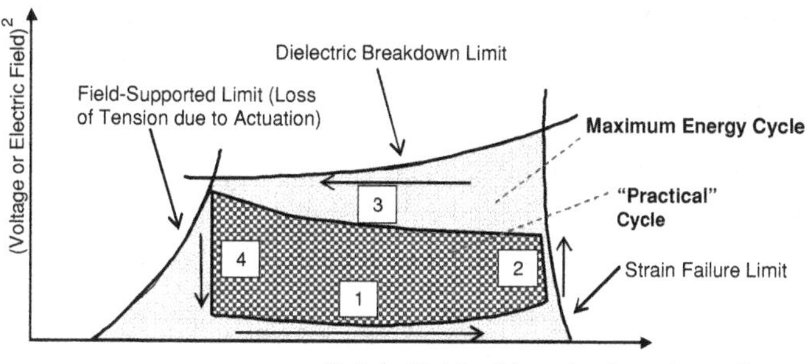

Fig. 3.3 Performance limits and energy cycles

(the film area divided by the initial film area A_0). The fact that the second of the equations of state reduces to the common equation for a capacitor is not surprising given the structure of the functional element. A dielectric elastomer generator is commonly and correctly described as a "stretchable capacitor." In the case of uniform biaxial stretching, $\gamma = \lambda^2$, but as we will see just below, this equation can also be used in the case of nonuniform biaxial stretching where γ is the state variable. Note also that the first equation is the same as that describing the actuation performance of a dielectric elastomer. Deformation of the polymer in response to applied voltages exists in both actuator and generator modes. In fact, because of this dual nature, a dielectric elastomer can also function as a variable stiffness device [7].

Writing simple equations for the amount of energy that can be extracted is not easy due to the highly nonlinear elastic behavior and the complex interactions with the energy source. If we focus on Eq. 3.2, by assuming that a given amount of stretch can be imposed on the film it is easier to see how a stretchable capacitor generator functions and how certain material and operational parameters affect the amount of energy generated.

While we cannot immediately discern the maximum amount of energy that can be produced from a given volume of material by this simplification, we can determine the energy output for certain operational cycles. There are four basic steps in the simplest operational cycles:

(1) the film is stretched by tensile forces to its maximum stretch state,
(2) a voltage or charge is applied to the film,
(3) the film relaxes from its internal elastic energy, and
(4) charge is removed from the film to return it to its initial state. Three common operational cycles are constant charge, constant voltage, and constant field. The names of these cycles refer to what occurs during step 3.

Figure 3.3 illustrates graphically an energy harvesting cycle. Note that the cycles must all be contained within the operational boundaries defined by the material limitations. The horizontal axis is a variable that represents the geometric change in the film (which is related to the change in capacitance). The vertical axis is the square of the voltage or electric field across the film. By choosing the correct variables, the energy that can be extracted for each cycle (not including losses) is proportional to the area enclosed by the cycle curve (e.g., capacitance versus the voltage squared).

The net amount of energy per unit volume of film that can be extracted for the constant charge, constant voltage, and constant field cycles are [8]

$$u_Q = \frac{1}{2}\varepsilon E_{\max}^2[(\gamma - 1/)\gamma^2] \tag{3.3}$$

$$u_V = \frac{1}{2}\varepsilon E_{\max}^2[(\gamma - 1/)\gamma^2] \tag{3.4}$$

$$u_E = \varepsilon E_{\max}^2 \ln(\gamma), \tag{3.5}$$

where $E_{\max}$ is the maximum field that is applied during the cycle. These equations can show the benefits of one cycle compared with another for different stretch conditions if we select materials based on the maximum field level. It is possible to implement cycles that can exceed these energy outputs by more closely approaching the material performance limits or including lower losses. Electrical losses result from resistive losses in the electrodes and leakage losses across the film. These losses can be modeled by adding series and parallel resistances to the capacitor representing the dielectric elastomer element. Indeed, the dielectric element can be modeled as a capacitance. There are additional electrical losses in the harvesting circuit and any storage or transmission systems. Graf et al. [9] modeled these losses by making simplifying assumptions such as constant polymer material conductivity and electrode resistance. In reality, these parameters will themselves be nonlinear functions of stretch and electric field, not to mention environmental effects such as temperature or humidity. Mechanical losses include viscoelastic losses in the polymer and electrodes as well as those in any mechanical transmission system that couples to the external driving loads. Here again, these loss mechanisms will tend to be nonlinear functions of stretch as well as sensitive to environmental conditions. Note also that, as shown in Eq. 3.1, the dielectric elastomer still experiences actuation forces when a voltage is applied and so the forces applied on the polymer by the external driving load will be affected by the energy harvesting cycle.

In many applications it is desirable to achieve the maximum output for a given amount of polymer material. However, in other applications maximizing the efficiency of the energy conversion will be desired. An analysis of efficiency must include all loss mechanisms discussed above. But we also note that for high efficiency, as well as practical considerations in implementation of the mechanical transmission system and energy harvesting circuits, we desire not just a large net energy output, but a large energy output compared with the energy input. This requirement can easily be understood by realizing that there are certain fixed losses in any mechanical transmission or electrical circuit that do not scale with energy output (e.g., bearing friction, switching losses). With regard to the electrical losses, we can get a sense of this efficiency by again considering the three simple energy harvesting cycles.

The relative energy gain, $G = u/u_{in}$ for each cycle (again from Graf et al. [8]) is

$$G_Q = \gamma^2 - 1 \tag{3.6}$$

$$G_V = (\gamma^2 - 1)/\gamma^2 \tag{3.7}$$

$$G_E = 2\ln(\gamma) \tag{3.8}$$

While the relative benefits of each cycle depend on the stretch ratio, for all cycles, a larger stretch ratio is beneficial. A very small stretch ratio suggests that it would be difficult to efficiently harvest the energy. For example, with a stretch ratio of 1.1, the constant voltage cycle has an energy gain of 0.17. For each joule of electrical energy applied to the polymer we would get out 1.17 J. Even a fairly efficient energy harvesting circuit that, let us say, only lost 0.1 J of energy, would only harvest 0.07 J and thus have an effective harvesting efficiency of 0.07/0.17 = 41%.

The overall efficiency is further reduced by additional losses as well as the fact that the system does not fully couple to the mechanical load. This coupling depends on several factors, including the stiffness of the material (discussed in more detail in the following section on material selection).

3.2.2 Materials

The performance of a material for dielectric elastomer generators depends on a combination of electrical and mechanical properties. From the simplified analysis of energy harvesting presented above, it can be seen that it is generally desirable to have a material that has high dielectric breakdown strength and high permittivity (dielectric constant). To minimize losses, it is desired to select a material with low leakage or, to a lesser extent, dielectric relaxation losses. The importance of leakage depends on the frequency of operation. A vibrational energy harvesting system might operate at more than 100 Hz while an ocean wave power harvesting system might operate at less than 0.1 Hz. Graf et al. [8] looks at the importance of leakage for a simplified model assuming constant leakage. More precise assessment of the importance of leakage would require a model of the energy harvesting circuit as well as correction for environmental conditions. On the mechanical side, it is generally desirable to have a material that can sustain large stretch ratios. It is also desirable to minimize viscoelastic losses as well as creep and stress relaxation effects.

The question of selecting the best material stiffness is a more complex issue. At first glance it would seem that it is best to choose a material with a low stiffness so that smaller forces are needed to produce the desired polymer stretch so that the generator structure can be simpler and there are fewer mechanical losses. To borrow from the study of piezoelectrics, it is generally desirable to have a greater coupling constant k^2 = electrical energy out/mechanical energy in. Clearly, a stiffer material will require more mechanical input energy to deform. But a lower stiffness is not always the most efficient way to couple to an energy source. In general, it is well-known that it is often desired to have an impedance match, that is, there is maximum energy transfer when the impedance of the load matches that of the source (including any mechanical transmission that couples the two as well as the electromechanical effects of the energy harvesting circuit). In other cases, it is desirable to maximize the energy transfer by operating at a resonant condition. In this case, the stiffness of the generator would be designed to make the resonant frequency close to the frequency of the driving force. Note that the impedance of the polymer is affected by the energy harvesting circuit. While this sensitivity makes the design of a system more complicated, it is also possible to tune the system for a better impedance match by controlling the energy takeoff. The effective stiffness seen by the driving mechanism depends on the amount of polymer material used as well as the geometry of the element. A short and wide transducer would appear stiffer than would a long narrow transducer with the same amount of film.

A low stiffness material may have some additional adverse effects. A low stiffness material will experience a loss of tension in the film in the field-supported region of operation at a lower electric field. Further, many soft materials would be more prone to pull-in failure due to mechanical instabilities resulting from film defects or thinner film regions (a source of dielectric failure in softer insulating films) [10].

The most common candidate materials considered for dielectric elastomer generators, just as with dielectric elastomer actuators, are those based on commercial formulations of acrylics and silicones [11]. These materials have a favorable combination of high dielectric breakdown strength, high elongation, and relatively low mechanical and electrical losses. As noted above, acrylics have greater breakdown strength at high prestrain. Ha et al. [12] report on acrylics that have been modified so as to not require high prestrain. Such materials can more easily be formed into a variety of transducer configurations. Other materials under development by researchers include SEBS and acrylonitrile rubbers as well as polyurethane-based polymers [3].

It is important to note that the best choice of material may not be just that which is capable of the greatest energy density but may also consider economic considerations. The effect of economic factors is more critical in large-scale energy harvesting (such as ocean wave power, as will be discussed below). Koh et al. [6] use their nonlinear material model to show how under some operating conditions natural rubber can outperform 3 M VHB acrylic. While this analysis claims similar maximum energy density for the two materials (which may not be a valid assumption under real-world conditions), it does not even consider the much lower cost of natural rubber.

In recognition of the advantages of high permittivity in achieving greater energy density, many researchers have experimented with adding particulates to elastomers in order to increase the permittivity. Generally, this approach increases permittivity at the expense of breakdown strength and leakage and so the net effect on energy density is detrimental. Recently, Kofod et al. [13] have shown that certain nanoparticles can increase the dielectric constant without such adverse effects. Whether such material improvements are beneficial may again depend on economic issues. We also note that applications designed to recharge a battery may benefit from a high permittivity material that can operate at relatively lower voltages. Large-scale power generation has no such requirements and, in cases where power is designed to feed a high-voltage transmission line, it is preferable to operate at the high voltages commonly used in transmission lines.

We have thus far considered the dielectric material only. The overall performance of a dielectric generator must also consider the electrode material that coats the surfaces of the films. In general, it is desired to make the electrode as compliant as possible. Because dielectric elastomers typically operate at high voltage and low current conditions, it is acceptable to use relatively high resistance materials, compared with piezoelectrics, for example. Electrode materials for dielectric elastomers typically include various carbon particles in polymer binders or patterned or corrugated metal coatings. A good overview of electrodes may be found in Ref. [14]. Most work on electrodes has been oriented toward actuation. For generation, the requirements are similar except that the materials may have to undergo even larger strains without creating a loss of conductivity or causing mechanical damage to themselves or the underlying polymer as will be discussed in Sect. 3.4.

Recently, silicone dielectric elastomer material already coated with compliant electrode material (corrugated silver) was introduced to the market [15]. We also note that the 3 M VHB acrylic (uncoated dielectric elastomer) is also available in large quantities. That such materials can be manufactured in large-scale roll-to-roll operations helps support the notion of the feasibility of large-scale power generation.

3.2.3 Transducer Configurations

The basic operational element of Fig. 3.1 must be incorporated into a transducer or structure that allows the stretching of the film to be coupled to the forces that cause stretching. Kornbluh [16] surveys a variety of configurations for actuators. These same configurations can also be applied to generators. Figure 3.4 shows several important configurations, many of which have been utilized in the application examples in the following section.

The selection of the best configuration depends on many factors, including the type of driving force and mechanical transmission, operating strain, total amount of film needed, and the desired form factor. It is desirable to have the boundary conditions impose a uniform strain over the entire range of operation. The examples in Fig. 3.4 come close to this ideal.

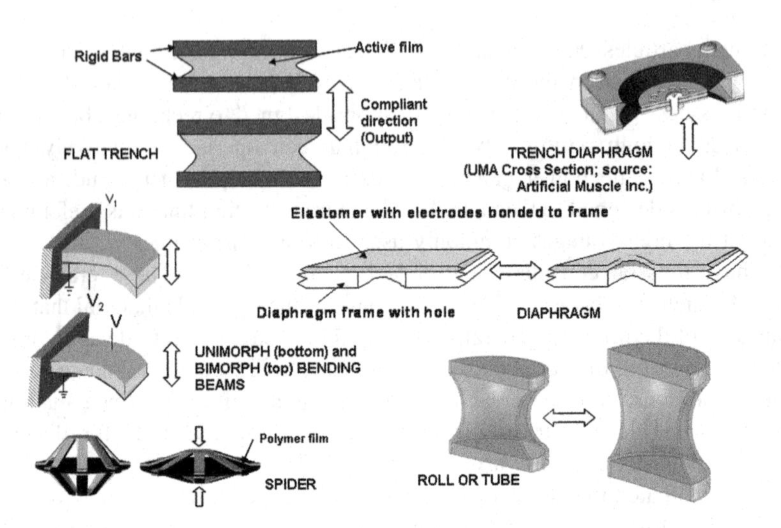

Fig. 3.4 Configurations for dielectric elastomer generators

3.2.4 Energy Harvesting Circuits

We have illustrated above how, in order to extract energy from a dielectric elastomer, the dielectric elastomer must be connected to an energy harvesting circuit. The energy harvesting circuit manages the introduction and extraction of charge to and from the film as well as the transfer of the charge to the electrical load (be it for immediate use or storage). Simple energy harvesting cycles, such as the constant voltage cycle, can use relatively simple circuits. Circuits that use more complex cycles that attempt to more fully approach the performance boundaries of the material are possible as well. However, such circuits may require microprocessor control or other means of adapting to highly nonlinear behavior and the often-unpredictable nature of harvestable energy sources (e.g., waves, footsteps). When the mechanical driving force is unpredictable, such circuits might also benefit from sensor feedback that indicates the state of stretch or other state variables. The practicality of implementing such circuits depends on the particular application and whether or not it is critical to maximize the energy output for a given amount of material.

The basic requirements of dielectric elastomer energy harvesting circuits are fairly similar to those of piezoelectric energy harvesters. There is extensive literature on the optimal design of such circuits for a variety of applications (e.g., Rupp et al. [17]). Further, there are commercially available devices (e.g., from Mide Technology Corporation, Medford, Massachusetts, USA) and even an integrated circuit chip (available from Linear Technology Corporation, Milpitas, California, USA). Circuits for dielectric elastomers differ in one fundamental way: In order to harvest energy from dielectric elastomers, a charge must first be placed on the film. Additional differences include the greater nonlinearity of the dielectric elastomer due to high-strain effects and the fact that dielectric elastomers typically operate at higher maximum voltages (thousands rather than hundreds of volts). This latter point is important because it affects the choice of available electrical components. In general, there is a dearth of commercially available switches and diodes that operate at the high voltage, low leakage current conditions needed for efficient operation (although this situation is changing in recent years with the introduction of higher voltage low-leakage transistors).

We noted that, unlike piezoelectric systems, dielectric elastomer generators must initially place a charge on the polymer. Typically, this charge is applied at high voltage in order to maximize energy output. While this charge may be applied with a power supply, such an approach can be more inefficient and costly. It is possible to design the energy harvesting circuit so that it is "self-priming," that is, a small amount of energy can initially be applied at low voltage and the voltage and energy output can be increased with each cycle up to the maximum operating conditions. Circuits capable of reaching operating voltages in the kilovolt range form initial charges of 10 V or less (such as can be supplied by the battery energy storage media) have been shown by the authors and others (e.g., McKay et al. [18]).

3.3 Unique Capabilities of Dielectric Elastomers for Energy Harvesting

3.3.1 Unique Capabilities and Comparison with Other Technologies

We have already touched on some of the unique properties of dielectric elastomers and the implications for energy harvesting. These unique properties and their implications are as follows:

- High energy density—low mass and bulk, ability to use simple and efficient direct drive transmission.
- High efficiency over a range of frequencies—high efficiency in the presence of unpredictable driving forces (particularly low frequencies).

Low (or selectable) compliance—can couple well to low-force high-strain loads such as those from human activity or wind and waves

- Large-area low-cost films with no toxic materials—enables economic capture of highly distributed diffuse power sources for large-scale power generation.
- Choice of material to meet system requirements including wide-range environmental tolerance.

Table 3.1 below quantifies some of these properties and compares them with common power generation technologies.

Table 3.1 Comparison of dielectric elastomer and other power generation technologies (adapted from Pelrine and Prahlad [19] and Jean-Mistral et al. [20])

Technology	Typical stiffness	Maximum specific energy density (J/g)	Typical max. efficiency (%)	Comments
Dielectric elastomer EAP	0.1–10 MPa	0.4 (0.05 for long lifetime operation)	>50	Low-cost materials, low stiffness size scalable
Electromagnetic	NA	0.004	<20	Low-energy density, constant-frequency electromagnetic generators can have much higher efficiency, but needs variable frequency transmission for higher efficiency over a range of frequencies
Piezoelectric ceramic	50–100 GPa	0.01	>50	Requires significant additional mass for support or motion amplification, expensive (and often toxic) materials

Other electronic (electric field responsive) electroactive polymers such as those based on copolymers of PVDF can, in principle, have energy densities similar to dielectric elastomers. However, this power output generation at these levels of output has not yet been experimentally demonstrated (e.g., Liu [21], Jean-Mistral et al. [20]). Jean-Mistral et al. also note that wet (ionic) electroactive polymers such as conductive polymers and IPMC have not shown the capacity for large energy densities. Piezoelectric polymers, including both PVDF and composites that include piezoelectric ceramics have similarly not shown high energy density. Further, these materials are generally more expensive than dielectric elastomers and cannot yet be readily produced into large-area films needed for large-scale power production. In general, the literature does not show examples of direct drive energy harvesting devices with energy outputs as large as those already achieved with dielectric elastomers (as revealed in the following sections).

There are a great many potential applications that can take advantage of the benefits of dielectric elastomer generators. These applications might be classified along various axes such as size and amount of energy output (e.g., power a small device or provide commercial-scale energy to the power grid), frequency of energy source (e.g., human activity such as pushing a button or high-frequency vibrations), type of energy source (e.g., point source vs. distributed), transducer type (e.g., diaphragm or roll), mechanical coupling from load (e.g., proof mass vs. fluid, direct or rotary coupling), and purpose of energy harvesting (e.g., power remote sensors, save energy, or cost of batteries). Table 3.2 highlights the potential benefits of dielectric elastomers for several categories of energy sources.

Table 3.2 Advantages of dielectric elastomers for various energy sources

Generator application	Competing technology	Dielectric elastomer potential advantage	Dielectric elastomer potential challenge
Human activity(e.g., heel strike or knee brace)	Electromagnetics, piezoelectrics	High energy density and low stiffness allows good load matching and eliminates much mechanical complexity, mass, and bulk	Electronics more complex than electromagnetics
Enviromental sources (waves, water flow, and wind)	Electromagnetics	Good matching to load; low cost materials allows for large-scale and highly distributed energy harvesting	Long lifetimes for large film areas, electronics cost
Fuel (or other heat sources) engine-driven generators	Electromagnetics	Higher energy density, lower cost good low-speed performance, higher temperature performance	Electronics cost and weight (very small engines), lifetime
Parasitic energy harvesting (vibrations) for remote sensors or other device	Electromagnetics, piezoelectrics	Good load matching to some available energy sources enables simpler designs; lower cost	Electronics cost an issue for some applications, high efficiency at low duty cycles

The following sections give examples of dielectric elastomer generators in several categories of energy source, size, transducer configuration, coupling type, and purpose.

3.3.2 Human Activity: Heel-Strike Generator

The proliferation of mobile electronics for the general public as well as warfighters or emergency first responders has put demands on the life of batteries and the need to simplify the logistics of recharging systems. Harvesting the energy of human activity can help. Starner's seminal work on wearable computers [22] surveys many potential sources of human power. Figure 3.5 shows several sources of human kinetic activity that can produce significant amounts of power. Several of these power sources are under development by researchers (e.g., Rome et al. [23], Alexander [24], and Kuo et al. [25]).

The authors have developed a "heel-strike generator" that can be located in a normal shoe or boot. The compression of the heel during normal walking was selected as the means of harvesting power from human activity because it does not add any physical burden to the wearer. Further, proper tuning of the amount of energy absorption at the heel could actually increase the comfort or walking efficiency of the wearer by absorbing and returning the optimal amount of energy

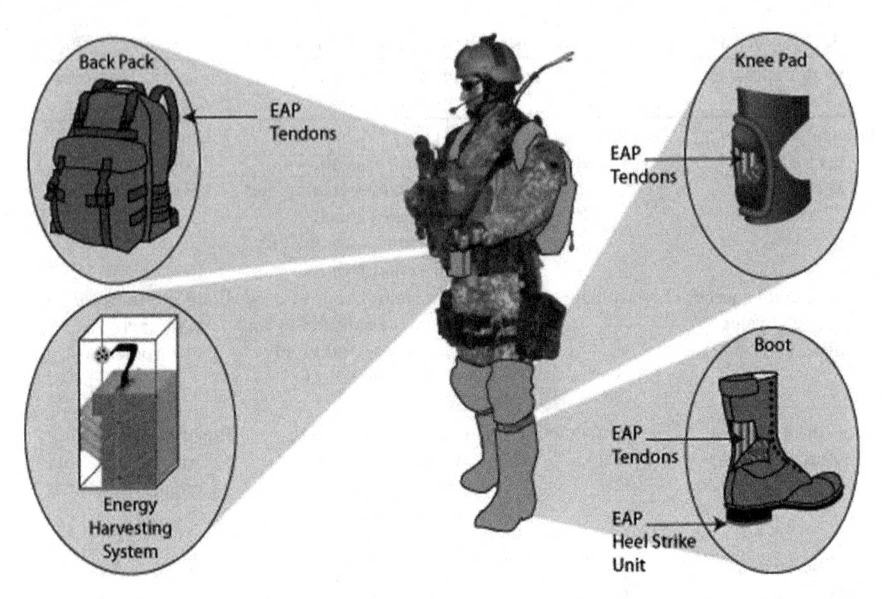

Fig. 3.5 Notional dielectric elastomer devices that can harvest human kinetic energy. Kinetic energy is available from vertical motion and from the negative work done at the ankles, knees, and hips. Energy harvesting devices can be strategically located in equipage to selectively harvest this energy in order to provide power to electronic systems. This arrangement will not hamper mobility and should actually reduce fatigue. [*Source* adapted from Infoscitex Corp. and SRI International]

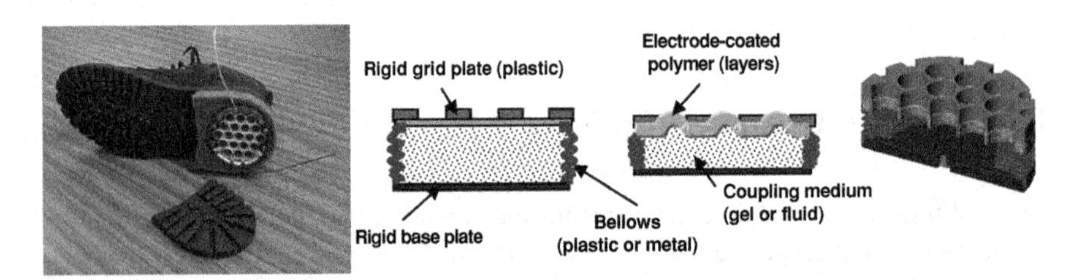

Fig. 3.6 Heel-strike generator based on dielectric elastomer; photo of device installed in boot (*left*), cross section of device (*right*)

per step. This device, shown in Fig. 3.6, was capable of producing an electrical output of 0.8 J per step, or about 1 W while walking. The diaphragms in this device consisted of 20 stacked layers of dielectric elastomer films. While intended primarily for battery charging, the device was also able to directly power night vision goggles. This device used prestrained VHB 4910 acrylic and performed with a maximum energy density of about 0.3 J/g. The generator was a diaphragm type that used a fluid (or gel) coupling to transfer the compression of the heel into deflection of the diaphragm.

This power level far exceeds outputs demonstrated by many other more complex, more costly, and heavier heel-strike generators based on direct deformation of piezoelectric elements [26] as well as direct-drive electromagnetic devices, and therefore supports the claims of high efficiency and energy density that are possible based on dielectric elastomers. By means of comparison, one can roughly estimate the available energy per step as weight times the maximum deflection of the heel and get, for example, 2.4 J of available energy from a 80 kg person and a maximum of 3 mm deflection. Thus, our 0.8 J represents 33% overall efficiency.

The energy recovery from a heel strike is limited by the amount of heel deflection. There is potentially greater energy available from the "negative energy" of braking applied by our muscles to the knee during normal walking. That is, we use energy to counteract dynamic forces trying to bend the knee. Winter [27] reports that there is up to 21 W of negative input energy available from this motion. Like the heel-strike, harvesting such energy should not place any additional burden on the wearer and can even make walking more efficient. Such a device is shown in Fig. 3.5 above.

3.3.3 Environmental Sources: Wave Energy Harvesting

New clean and renewable sources of electric power are critical as the world moves toward a more secure and sustainable energy future. Wind and solar power suffer from the lack of on-demand availability and great daily and seasonable fluctuations. Commercial-scale wind and solar energy projects require the construction of large structures that have adverse visual and noise impacts to neighbors and are hazardous to wildlife, especially birds. Ocean wave power has the potential to produce clean renewable energy in an environmentally sound manner that offers greater reliability than solar or wind and lower visual and auditory impact than wind. The Electric Power Research Institute estimates that wave energy could meet 10% of total worldwide electric demand [28]. Unlike other sources of ocean energy such as tidal or ocean thermal, wave power is widely available. Further, it tends to be available near many centers of population and industry.

Despite these benefits, widespread adoption of wave power harvesting is hampered by certain economic and logistical factors. For instance, the primary converter structure of conventional ocean wave power harvesting systems must be over-engineered to deal with high sea events, and, as a result, are very expensive. Similarly, efficient power take-off systems (the structure and transmission systems needed to convert the hydrodynamic energy into electrical power) are typically highly complex and expensive. Dielectric elastomers can potentially address these issues by enabling a simple low-cost power take-off system.

SRI developed and demonstrated the use of dielectric elastomers for harvesting the energy of ocean waves. This work included two sea trials in which SRI deployed a complete energy harvesting system at sea. Figure 3.7 shows the system for the first sea trial conducted in Tampa Bay near St. Petersburg, Florida.

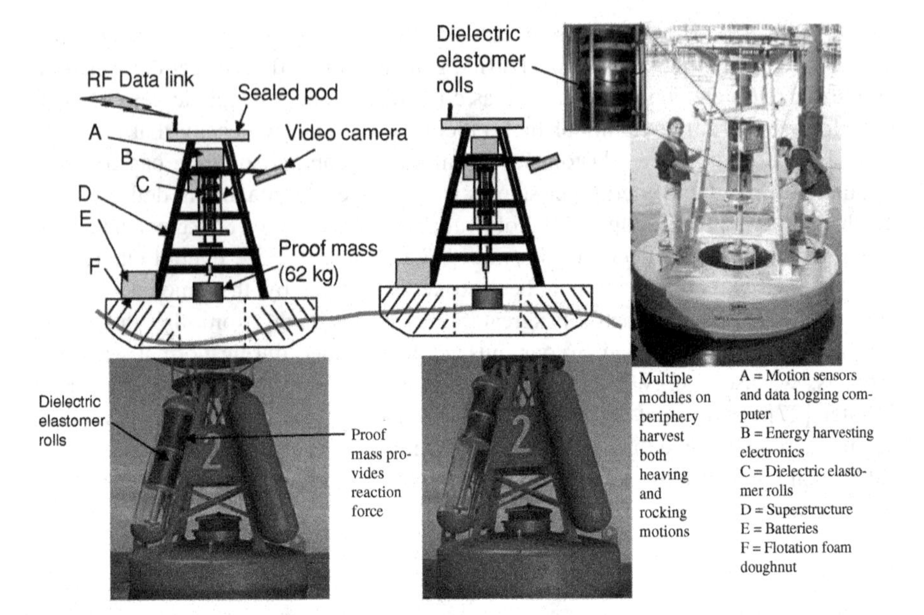

Fig. 3.7 Dielectric elastomer ocean wave power generator based on a proof mass, system tested at sea (*top*) and CAD model of a final system that might be used for supplying power to navigation and scientific buoys (*bottom*)

This generator was based on a suspended, "proof-mass" approach that used concatenated roll transducers.

The dielectric elastomer generator unit has a cylindrical shape (diameter of 40 cm and height of 1.2 m) as shown in Fig. 3.4. Two dielectric elastomer elements were installed into the generator module. Each element consisted of an active amount of dielectric elastomer film of 150 g, which was wrapped to form a roll with a diameter of 30 cm and a length of 20 cm (active length in the stretched condition). The maximum measured electrical output capacity, verified in laboratory tests, was 12 J for one cycle of operation (0.08 J/g). The mechanical structure that stretched and contracted the dielectric elastomer rolls was quite simple. A mass of 62 kg was attached to the rolls. The inertial force of the mass in response to the wave-induced motion of the buoy causes the stretching and contraction of the rolls.

The dielectric elastomer rolls remained operational during the sea trails despite experiencing temperatures of up to 55°C inside the transparent enclosure of the generator prototype (the enclosure was transparent for observational purposes only) and ambient humidity occasionally approaching 100%.

There was little wave activity during these sea trials. The largest waves experienced during the tests were on the order of 10 cm peak-to-peak wave height. Despite the low wave activity, the generator was shown to produce net energy per cycle. Even with the small wave heights of 10 cm, we were able to generate a peak power of

Fig. 3.8 Dielectric elastomer ocean wave power generator based on a an articulated multibody system, individual roll transducer (*top left*), concatenated rolls in a generator module (*bottom left*), and buoy at sea trial site (*right*)

1.2 W with an average power of 0.25 W. While this power is a very modest amount, it does attest to the efficiency of the energy harvesting system. Further, we should note that these measurements were made with a small bias voltage of 2,000 V applied to the dielectric elastomer. Raising the bias voltage to 6,000 V (a value sustainable for at least short-term operation with the acrylic elastomer material) would have produced a peak power of 11 W and an average power of 2.2 W. While this amount of power still seems quite modest, it is easy to see how more significant wave heights would produce more useful amounts of power. The power output is roughly related to the square of the wave height. Thus, at wave heights of just 0.5 m, with more dielectric elastomer material, a generator operating at the same demonstrated level of performance would produce an average power of 50 W—more than enough to supply the power needs of a navigation buoy.

As a self-contained system, the proof mass approach is well-suited to producing power for signal or weather or oceanographic monitoring buoys. However, such an approach may not be practical for large-scale power generation designed to produce grid-level power outputs due to the large proof mass that would be needed. To that end, SRI developed a proof-of-principle system that was based on the direct use of hydrodynamic energy to mechanically stretch and contract the dielectric elastomer. This system is shown in Fig. 3.8. For logistical convenience, the system was based on the same oceanographic buoy platform as the proof-mass system. We note that an optimum system would likely not use such a platform.

This system was tested at sea in the Pacific Ocean near Santa Cruz, California. The device produced an output of more than 25 J in laboratory testing. It used about 220 g of active dielectric elastomer material for a corresponding energy density of more than 0.1 J/g. At sea, the buoy was tested at about half this energy density. The energy harvesting circuit used in this sea trial was 78% efficient; that is, it harvested 78% of the expected energy for the particular energy harvesting cycle (roughly speaking, a constant voltage harvesting cycle but with a constant charge portion at the beginning of contraction). The harvested energy was stepped-down to 12 V and used to charge a battery.

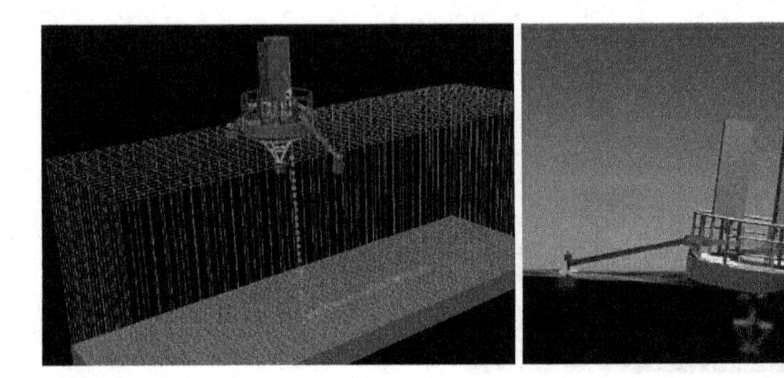

Fig. 3.9 Physical modeling of wave/buoy interaction; stability and mooring analysis of buoy with outriggers (*left*), dynamic response to waves (*right*)

Ocean wave power energy harvesting also serves as an example of the complexities of modeling an entire energy harvesting system. Physical modeling of the overall system is critical both for guiding design and making economically driven design and operational decisions.

For example, we desired to model both the hydrodynamic stability of our buoy and the energy output for various design configurations and wave conditions. No single commercially available modeling tool was found to be sufficient for all aspects of this modeling. For example, Orcaflex (Orcina, Cumbria, UK) is a simulation and modeling package specifically designed for evaluating the designs of off-shore vessels, platforms, and buoys. It can only model user-provided shapes as a collection of approximate cylinders and does not interface with common CAD design software, such as that which might be used to model the load coupling and transmission system. Therefore, we developed our own approach based on combining several modeling systems. The basic design was modeled using SolidWorks (Dassault Systèmes SolidWorks Corp., Concord, Massachusetts, USA). Motion analysis was implemented using the COSMOSMotion package. However, we used simplifying assumption to estimate hydrodynamic effects. Hydrostatic buoyancy effects including the overall system stability and mooring system was modeled using 3D Studio Max (Autodesk Media and Entertainment, Montreal, Quebec, Canada). The dynamic simulation engine of this package provides capability to calculate buoyancy forces based on complex polygonal geometry, and the interaction with the fluid is bidirectional, where the rigid objects disturb the free surface, and the surface influences the motion of the rigid objects. Figure 3.9 shows a wire-frame rendered model and full dynamic rendering of the energy harvesting buoy discussed above.

In this analysis, the dielectric elastomer transducer and energy harvesting circuits were modeled with highly simplified, experimentally validated lumped parameter models that did not include the interactions resulting from Eq. 3.1. Such models also did not include all of the nonlinear effects detailed in Sect. 3.2 above. The circuits were also modeled separately using more specialized circuit modeling

Fig. 3.10 Highly modular distributed ocean wave energy harvesting system based on dielectric elastomer transducers

software such as PSPICE (Cadence Design Systems, Inc., San Jose, California, USA). The system design could have benefited from a single integrated model.

The ocean wave energy harvesting buoys described above were proof-of-principle systems whose structure and mechanics were not optimized for maximum efficiency or economic benefit. In particular, many environmental energy sources are highly distributed, that is, harvesting large amounts of power requires that the system be spread out over a large area. This issue has been a limitation for wind and wave power, as well as solar. However, because of the low cost and simplicity inherent in using dielectric elastomer materials for energy harvesting, fundamentally new system designs can be enabled. Figure 3.10 shows a conceptual design of such a generator. The basic harvesting element is similar to that used in the single buoy device of Fig. 3.9, but here it is built into a highly modular system that can be easily assembled and transported and whose size can be tailored to the prevailing wave conditions (e.g., open ocean deep water waves or waves that might hit an existing seawall or breakwater).

Flow energy from microhydro sources in rivers or from tidal or ocean current flow can also be another significant clean and renewable energy source. There are expected to be a great many places suitable for installation. Tests are being carried out to develop water mill-based generation systems, including a portable floating type that is suitable for use as a source of electricity in disaster areas or in emergency situations, as well as a source of electricity in agricultural regions or in mountains, where supplying electricity is challenging. Figure 3.11 shows a proof-of-principle water mill developed by HYPER DRIVE Corp. (Japan) that is 30 cm in diameter. A small water pump (280 ml/sec flow rate) was used to move water that spun the waterwheel. The waterwheel was attached to a crankshaft with a push rod that was then attached to a diaphragm-trench type of transducer. Each turn of the wheel produced 35 mJ of electricity. While this amount of energy is quite small, it is limited by the small transducer. Estimates suggest that a water mill generator system with water mills of 80 cm in diameter could generate approximately 5.4–6 J of electrical energy per revolution of the wheel.

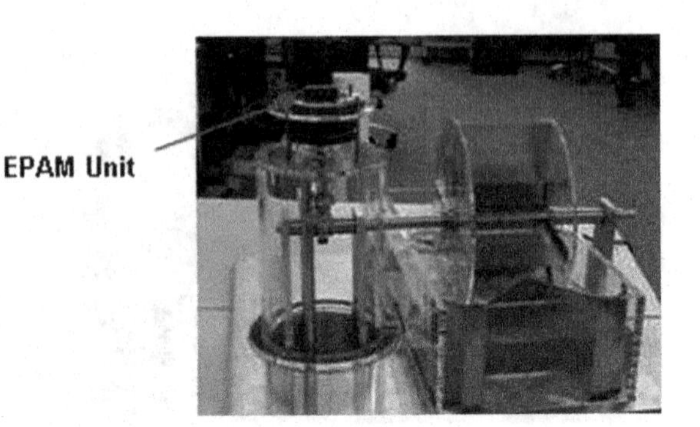

Fig. 3.11 Water mill generator using a dielectric elastomer transducer (*upper left* of photo) [*Source* HYPER DRIVE Corp.]

While this particular design relies on a conventional rotary wheel, dielectric elastomers can still help reduce the cost of and improve the efficiency of energy harvesting systems that need to operate over a wide range of speeds. To a first order, the efficiency of dielectric elastomers, unlike rotary electromagnetic systems does not depend on the operational frequency. Future systems based on flowing water or air might be even simpler and rely on undulating flaglike structures [29]. The spread of the use of these generators will help in the expansion of environments where electricity can be used, contributing to the support of agriculture, forestry, and other industries.

The water mill demonstrates how a dielectric elastomer generator can be used with a rotary energy source. Such a fuel or heat source-powered engine-generator could, in principle, offer advantages of simplicity or efficiency in converting chemical fuel into electrical energy. But such an approach does not fully exploit the potential advantages of dielectric elastomers. The authors have further simplified a dielectric elastomer heat engine by making the cylinder itself out of dielectric elastomer. In other words, expanding gases directly drive the expansion of the dielectric elastomer. In addition to minimizing mass and structure, this approach also allows for greater efficiency of a small engine because there are fewer losses from fuel leakage or friction of sliding seals, less wear, and potentially less heat loss for the same mass since polymer is a better thermal insulator. We have shown that a polymer cylinder can indeed sustain the temperature of combustion. A polymer engine was shown to be capable of 11% fuel-to-mechanical efficiency—a good value for a small (<20 W) engine. Figure 3.12 shows the expansion of a rolled dielectric elastomer actuator due to combustion of butane. A small, milliwatt-level amount of electrical power was generated with this proof-of-principle device.

In addition to rolls, we have demonstrated that diaphragms and tubes can also be used as cylinders of a polymer engine. In addition to combustion of fuel as an energy source, such engines might harvest solar energy. Such a simple engine design can enable unique energy harvesting systems based on many small distributed polymer cylinders and might form the basis of a new type of solar thermal system.

Fig. 3.12 Dielectric elastomer cylinder of a "polymer engine" undergoing 23% linear expansion during internal combustion

3.4 Lifetime Issues

We have so far ignored a major issue in the practical use of dielectric elastomer power generation for many applications—lifetime. While some applications might require only a few cycles of operation, in many applications it will be necessary to operate for many millions of cycles. Applications that harvest higher-frequency vibrations or harvest energy from fast rotating or oscillating engines or machinery may need to survive for billions of cycles.

The mechanical fatigue of rubbery materials has been extensively studied (e.g., Mars and Fatemi [30]). Similarly, the dielectric breakdown fatigue of insulators has also been extensively studied (e.g., Zakrevskii [31]). Yet, no studies combine the two effects. Further, a lifetime model must consider the interaction between electrical and mechanical effects resulting from the electromechanical forces due to charges on the surface electrodes as well as factors resulting from the mechanical properties of the electrodes themselves. The situation is further complicated by the fact that energy harvesting systems often need to operate in a range of environmental conditions and must respond to non-repeating mechanical (and resulting electrical) loading conditions. Further, all transducer configurations have boundary conditions and other sources of nonuniform behavior. It is also possible that the electrodes do not evenly distribute charge or themselves impose stress concentrations on the polymer film. While we do not propose to develop a unified theory here, we note that our experience shows that the expected lifetime will be adversely affected by both mechanical and electrical factors. In other words, the expected lifetime will be substantially less than that suggested by purely mechanical or electrical measurements alone.

There have been few studies on the lifetime of dielectric elastomer transducers and fewer still that consider lifetime of dielectric elastomer generators. Plante and Dubowsky [32] studied failures in acrylic materials and identified factors to predict performance limitations. Kornbluh et al. [33] did report some generator lifetime results, which will be highlighted here. The requirement for long lifetime can have

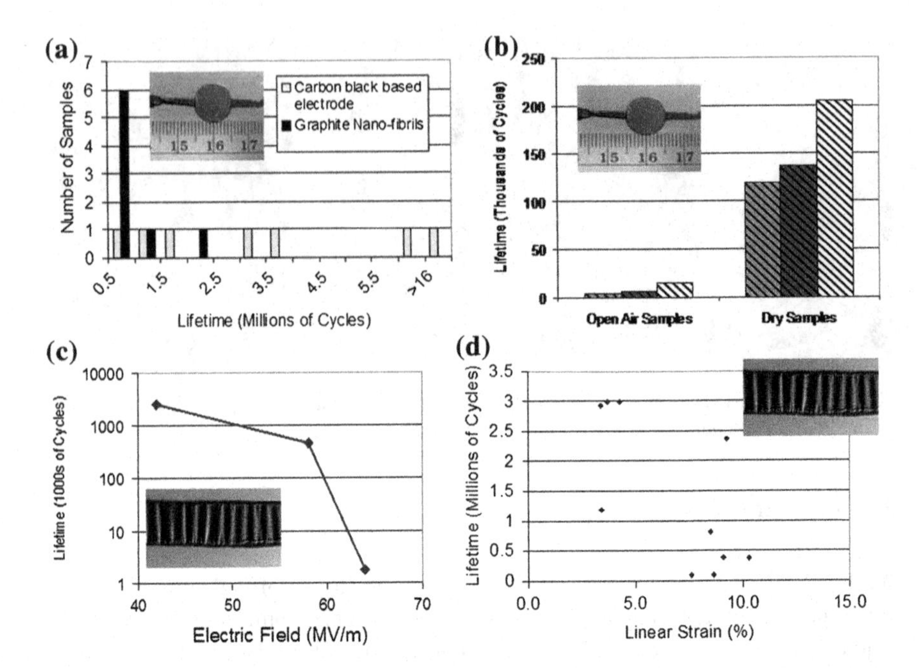

Fig. 3.13 Effect of electrode type, humidity, maximum operating field and strain on the lifetime of dielectric elastomer transducers **a Electrodes: distribution of circular high strain actuators operated with different electrodes formulations** (3M VHB 4910 film, ~ 50% RH, 300% × 300% prestrain, actuation real strain 30–40% at 5 Hz, Max field 140 MV/min). **b Humidity: difference in high-field lifetimes for six circular actuators, three in open air and three in a dry environment** (VHB 4910, 300% × 300% prestrain, 1Hz, Max field 140 MV/min). **c Electric field: average life time versus electric field of high-humidity actuators** (VHB 4910, 100% RH, 300% × 300% prestrain, 5% uniaxial strain at 5 Hz). **d Strain: lifetime of ten actuators with differing strain operated at high humidity** (VB 4910, 100% RH, 300% × 300% prestrain, uniaxial strain at 5 Hz)

a profound effect on the performance of dielectric elastomer generators. For example, the diaphragm transducer of the heel-strike generator was able to operate with an energy density of 0.3 J/g. The device was not tested for lifetime at this energy density. On the other hand, a draped roll transducer, similar to that used in the wave power harvester of Fig. 3.8, was able to operate at 0.05 J/g for more than 1 million cycles with materials and operating cycles (~ 100% areal strain) similar to the heel-strike transducer. This energy density is nearly an order of magnitude lower. As this comparison illustrates, the achievable energy density and lifetime are closely related—increasing one reduces the other.

Figure 3.13 shows the effect of several parameters on the lifetime of dielectric elastomer transducers made using the 3 M 4910 acrylic polymer. As suggested by studying Fig. 3.13, there are various material and operating factors that affect lifetimes. Electrodes should avoid field-concentrating components, operating humidity should be low, and maximum operating fields and strains should be low.

Fig. 3.14 Dielectric elastomer test samples submerged in saline solution

No doubt, further tuning of the materials and operating parameters will improve lifetime even beyond these levels.

In some generator applications it may not be possible or desirable to dry and seal the device. Nonetheless, data from dielectric elastomer actuator lifetime tests suggest that long lifetimes can still be achieved by a tradeoff in performance. For example, Fig. 3.14 shows operation of dielectric elastomer actuators submerged in salt water. In underwater operation, 6 out of 11 actuators survived for >10 million cycles with an electric field limited to 32 MV/m and approximately 2% strain (actuation strain). Operation while submerged in saline solution suggests the practicality of low-cost highly distributed ocean wave harvesters.

While these data are for acrylics, it is likely that similar effects apply to other materials. While we can use lifetime data to select the best materials and transducer designs, device design and systems analysis must consider tradeoffs between operating conditions such as electric field and stretch ratio. In some small devices, we can reduce operating conditions and still get the desired energy output by using more materials. In other cases, reducing operating conditions would result in a device that is too massive or cannot achieve the same energy output due to the economic considerations of using more material (and resulting need for greater structure to support greater amounts of material) as well as reduced electrical efficiency due to low stretch ratios.

Another important consideration is the amount of material that must be used to produce the needed power. For applications that require relatively small amounts of film, high lifetime can be readily achieved by using good quality film, properly selecting the electrode, and limiting the applied voltage. For example, in actuator mode, Artificial Muscle, Inc. (AMI—a Bayer Material Science Company, Sunnyvale, California) has developed dielectric films based on a non-acrylic material that it specifies for its X-Mode (planar) configuration (510 pF) to ten million cycles with 99% confidence at 25°C and 45% RH at strains of 5.9% at 36 MV/m. AMI also specifies to one million cycles with 97% confidence the same film at the accelerated aging condition of 65°C and 85% RH at strains of 5.9% at 36 MV/m. AMI has tested films to one billion cycles under some operating conditions [34]. Not surprisingly, AMI has found that reliability can be very sensitive to the manufacturing process. Some of the improvements in reliability that AMI has realized are due to enhancements in the control of the manufacturing process. These devices use only a square centimeter or so of active material. On a somewhat larger scale, Danfoss PolyPower (Nordborg, Denmark) claims lifetimes of ten million cycles with its silicone roll actuators: Ref. [35] These actuators typically operate at less than 5% strain. Danfoss is developing materials (polymer film with corrugated metal electrodes) for larger strain applications [36], but there is no extensive lifetime information yet published.

One approach to dealing with the increased likelihood of failures in larger area films is to enable such failures to "self-heal." For example, Lam et al. [37] report self-healing using electrodes based on a thin coating of nanofibrils. Sufficiently thin metal electrodes, such as used by Danfoss' materials, can self-heal as well. In generator applications where the strain can be large, self-healing electrodes may not be sufficient since even small film defects can easily propagate over time. Thus, any type of self-healing or fault tolerance would likely need to include means for subdividing the active film into individual regions such that a failure in an individual region does not result in catastrophic device failure. Such an approach was taken in the ocean wave energy harvesting device described above. A total of 40 individual films were used to produce the 25 J of output. Failed films could be replaced or removed from operation with electrical switching. A full-scale system designed for feeding power to the grid could easily have thousands of individual film elements.

3.5 Summary and Conclusions

Dielectric elastomer generators are capable of good characteristics and performance, both in theory and in experiment and demonstration devices. These properties include high energy density and high efficiency. Devices such as the heel-strike generator and ocean wave power harvesters have also demonstrated the ability to produce simple low-compliance devices that can directly couple to the mechanical energy source. The ocean wave energy harvesters have shown the possibility of developing devices with large amounts of film that can produce significant amounts of power. No other direct drive smart material technology has produced as much energy per stroke as has been demonstrated with dielectric elastomer systems. Despite this potential and progress, several challenges remain.

While progress is being made, fully modeled systems that include the full range of electromechanical coupling effects and environmental sensitivities do not yet exist. Further, the necessary software tools to model or solve for the material behavior, nonlinear electrical effects, and complex interactions with the environment are not available.

Even if the appropriate models and model solving tools were readily available, the necessary material parameters are not well established—even for well-studied materials such as silicones and acrylics. The inability to fully model a system does not allow for a full exploration of design tradeoffs and optimizations. In particular, the effects of lifetime on the maximum material parameters of electric field and strain are not well-known.

Better modeling tools would not only allow for better design or material selection, but could help guide the development of new materials. Ideal materials for generators may have different properties than those that are ideal for actuators. For example, being able to undergo very high stretch ratios might be more important than a higher dielectric constant. Being able to operate well in a high-humidity environment might also prove to be an important advantage for some applications.

The design of energy harvesting circuits is another area that has opportunities for further development. In many cases there will need to be tradeoffs between circuit complexity (to get high efficiency) and simplicity or cost. Further, the optimal energy harvesting cycle cannot be implemented unless the material is well characterized and modeled. Large-scale energy harvesting systems might benefit from numerous simple energy harvesting circuits as opposed to more centralized and sophisticated circuits. Again, integrated modeling can help address this issue. In some cases, energy-harvesting circuitry could be too large and/or too expensive for a given application, negating many of the benefits of using dielectric elastomers. To date, there has been little market for transistors suited to the relatively high voltages used in electroactive polymer energy-harvesting circuits. As a result, few such transistors are available in the marketplace. As better high-voltage transistors become available and harvesting circuits are refined, the shortcomings of today's circuitry can be overcome.

Large-scale energy harvesting, such as those for harvesting wave or wind energy, in particular can benefit from the improved modeling and design tools. Such large-scale systems will also require some degree of fault tolerance. Fault tolerance strategies must be further developed.

With better modeling and design tools, as well as more fully characterized materials and new materials, dielectric elastomer power generation may prove to be well-suited for a wide range of energy harvesting applications from the small to the very large. Physically small applications will likely be first, because the technological and economic barriers are lower. In order to enable the physically large applications, such as wave power harvesting, we will need advances in large transducer fabrication, lifetime, the energy-harvesting circuitry, modeling and system engineering.

Acknowledgments The authors wish to thank their colleagues at SRI international whose efforts contributed to the work presented here. We would also like to thank the numerous clients and government funding agencies whose support over the past 20 years has enabled much of the work presented here. We would like to thank in particular Mr. Shuiji Yonmura and Mr. Mikio Waki of HYPER DRIVE Corp., a company that has generously supported our development of the ocean wave power harvesting systems. Infoscitex Corporation contributed valuable information on human kinetic energy harvesting through Mr. Jeremiah Slade.

References

1. Pelrine R, Kornbluh R, Pei Q, Joseph J (2000) High-speed electrically actuated elastomers with over 100% strain. Science 287(5454):836–839
2. Kornbluh R, Pelrine R, Pei Q, Oh S, Joseph J (2000) Ultrahigh strain response of field-actuated elastomeric polymers. Proc SPIE 3987:51–64
3. Brochu P, Pei Q (2010) Advances in dielectric elastomers for actuators and artificial muscles. Macromol Rapid Commun 31:10–36
4. Carpi F, DeRossi D, Kornbluh R, Pelrine R, Sommer-Larsen P (2008) Dielectric elastomers as electromechanical transducers. Fundamentals, materials, devices, models and applications of an emerging electroactive polymer technology. Elsevier Press, Amsterdam
5. Pelrine R, Kornbluh R, Eckerle J, Jeuck P, Oh S, Pei Q, Stanford S (2001) Dielectric elastomers: generator mode fundamentals and applications. Proc SPIE 4329:148–156
6. Koh SJA, Keplinger C, Li T, Bauer S, Suo Z (2011) Dielectric elastomer generators: how much energy can be converted. IEEE/ASME Trans Mechatron 16:3–41
7. Carpi F, DeRossi D, Kornbluh R, Pelrine R, Somer-Larsen P (2008) Dielectric elastomers as electromechanical transducers. fundamentals, materials, devices, models and applications of an emerging electroactive polymer technology. Elsevier Press, Amsterdam, pp 141–145
8. Graf C, Maas J, Schapeler D (2010) Energy harvesting cycles based on electro active polymers. Proc SPIE 7642:764217-1–764217-12
9. Graf C, Maas J, Schapeler D (2010) Optimized energy harvesting based on electro active polymers. (ICSD), 2010 10th IEEE international conference on solid dielectrics, pp 752–756
10. Pelrine R, Kornbluh R, Joseph J (1998) Electrostriction of polymer dielectrics with compliant electrodes as a means of actuation. Sens Actuators A Phys 64:74–85
11. Carpi F, DeRossi D, Kornbluh R, Pelrine R, Somer-Larsen P (2008) Dielectric elastomers as electromechanical transducers. Elsevier Press, Amsterdam, pp 33–42
12. Ha SM, Park IS, Wissler M, Pelrine R, Stanford S, Kim KJ, Kovacs G, Pei Q (2008) High electromechanical performance of electroelastomers based on interpenetrating polymer networks. Proc SPIE 6927:69272C1–69272C9
13. Kofod G, McCarthy DN, Stoyanov H, Kollosche M, Risse S, Ragusch H, Rychkov D, Dansachmuller M, Wache R (2010) Materials science on the nano-scale for improvements in actuation properties of dielectric elastomer actuators. Proc SPIE 7642:76420J. doi:10.1117/12.847281
14. Carpi F, DeRossi D, Kornbluh R, Pelrine R, Somer-Larsen P (2008) Dielectric elastomers as electromechanical transducers. fundamentals, materials, devices, models and applications of an emerging electroactive polymer technology. Elsevier Press, Amsterdam Chapter 7
15. Benslimane M, Kiil H-E, Tryson MJ (2010) Electromechanical properties of novel large strain PolyPower film and laminate components for DEAP actuator and sensor applications. Proc SPIE 7642:764231
16. Carpi F, DeRossi D, Kornbluh R, Pelrine R, Somer-Larsen P (2008) Dielectric elastomers as electromechanical transducers. fundamentals, materials, devices, models and applications of an emerging electroactive polymer technology. Elsevier Press, Amsterdam, pp 79–90
17. Rupp CJ, Dunn ML, Maute K (2010) Analysis of piezoelectric energy harvesting systems with non-linear circuits using the harmonic balance method. J Intell Mater Syst Str 2010 21:1383. originally published online 10 Sept 2010
18. McKay T, O'Brien B, Calius E, Anderson I (2010) Self-priming dielectric elastomer generators. Smart Mater Struct 19(5):055025
19. Carpi F, DeRossi D, Kornbluh R, Pelrine R, Somer-Larsen P (2008) Dielectric elastomers as electromechanical transducers. fundamentals, materials, devices, models and applications of an emerging electroactive polymer technology. Elsevier Press, Amsterdam, pp 146–155
20. Jean-Mistral C, Basrour S, Chaillout J-J (2010) Comparison of electroactive polymere for energy-scavenging applications. Smart Mater Struct 19(8):085012

21. Liu Y, Ren KL, Hofmann HF, Zhang Q (2005) Investigation of electrostrictive polymers for energy harvesting. IEEE Trans Ultrason Ferroelectr Freq Control 52(12):2411–2417
22. Starner T (1996) Human powered wearable computing. IBM Syst J 35(3):618–629
23. Rome LC, Flynn L, Goldman EM, Yoo TD (2005) Generating electricity while walking with loads. Science 309(5741):1725–1728
24. Alexander RM (2005) Models and the scaling of energy costs for locomotion. J Exp Biol 208:1645–1652
25. Kuo AD, Donelan JM, Ruina A (2005) Energetic consequences of walking like an inverted pendulum: step-to-step transitions. Exerc Sport Sci Rev 33:88–97
26. Paradiso JA, Starner T (2005) Energy scavenging for mobile and wireless electronics. IEEE Pervasive Comput 4(1):18–27
27. Winter DA (1983) Moments of force and mechanical power in jogging. J Biomech 16:91–97
28. EPRI (2005) Ocean tidal and wane energy, renewable energy technical assessment guide. TAG-RE 1010489
29. Prahlad H, Kornbluh R, Pelrine R, Stanford S, Eckerle J, Oh S (2005) Polymer power: dielectric elastomers and their applications in distributed actuation and power generation. In: Proceedings of ISSS 2005 international conference on smart materials structures and systems, pp SA-100–SA-107
30. Mars WV, Fatemi A (2002) A literature survey on fatigue analysis approaches for rubber. Int J Fatigue 24(9):949–961
31. Zakrevskii VA, Sudar NT, Zaopo A, Dubitsky YA (2003) Mechanism of electrical degradation and breakdown of insulating polymers. J Appl Phys 93:2135
32. Plante J-S, Dubowsky S (2006) Large-scale failure modes of dielectric elastomer actuators. Int J Solids Struct 43:7727–7751
33. Kornbluh R, Wong-Foy A, Pelrine R, Prahlad H, McCoy B (2010) Long-lifetime all-polymer artificial muscle transducers. Proceedings of 2010 MRS spring meeting, Symposium JJ
34. Rosenthal M, Biggs SJ. Personal communication, March (2010) and February (2011)
35. Thomsen B, Tryson M (2009) Highly accelerated stress testing (HAST) of DEAP actuators. Proc SPIE 7287:102–113
36. Mohamed B, Kiil H-E, Tryson MJ (2010) Electromechanical properties of novel large strain PolyPower film and laminate components for DEAP actuator and sensor applications. Proc SPIE 7642:764231
37. Lam T, Tran H, Yuan W, Yu Z, Ha SM, Kaner R, Pei Q (2008) Polyaniline nanofibers as a novel electrode material for fault-tolerant dielectric elastomer actuators. Proc SPIE 6927: 69270O-4

Chapter 4
Theory of Ionic Electroactive Polymers Capable of Contraction and Expansion–Contraction Cycles

Lenore Rasmussen

Abstract Electroactive polymers (EAPs) that bend, swell, ripple (first generation materials), and now contract with low electric input (new development) have been produced. Ras Labs also produces EAP materials that quickly contract and then expand, repeatedly, by reversing the polarity of the electric input. Using applied voltage step functions produces varying amounts of contraction, which has enormous potential. A combination of high and low voltages could produce gross and fine motor skills, respectively, with both large motor control and fine motor control (fine manipulation) within the same actuator unit. The mechanism of contraction is not well understood. Radionuclide-labeled experiments were conducted to follow the movement of electrolytes and water in these EAPs. In addition, other experiments were conducted to determine how and why contraction occurs. One of the biggest challenges in developing these actuators, however, is the electrode-EAP interface because of the pronounced movement of the EAP. Plasma treatment of the electrodes, along with other strategies, allows for the embedded electrodes and the EAP material of the actuator to work and move as a unit, with no detachment, by significantly improving the metal–polymer interface, analogous to nerves and tendons moving with muscles during movement.

Keywords Electroactive polymer · EAP · Contraction · Artificial muscle · Plasma treatment · Actuator

L. Rasmussen (✉)
Ras Labs, LLC, Intelligent Materials for Prosthetics and Automation,
Plasma Surface Modification Experiment,
US Department of Energy's Princeton Plasma Physics Laboratory at Princeton University,
Room L-127, 100 Stellarator Road,
Princeton, NJ 08543, USA
e-mail: rasmussl@raslabs.com

L. Rasmussen (ed.), *Electroactivity in Polymeric Materials*,
DOI: 10.1007/978-1-4614-0878-9_4,
© Springer Science+Business Media New York 2012

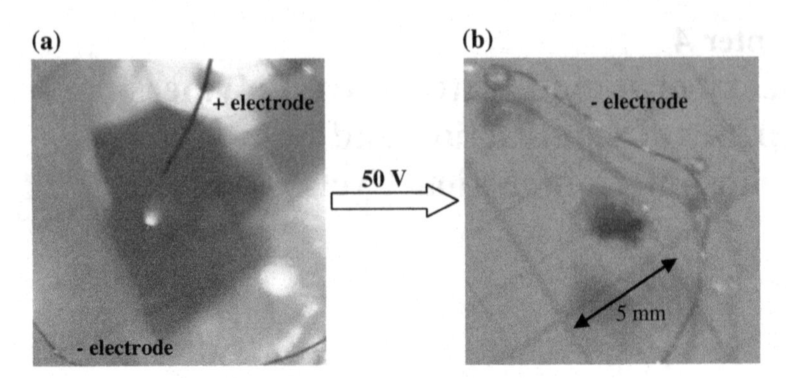

Fig. 4.1 Contraction of highly contractile EAP (**a**) Weight (t = 0) = 0.11 g (**b**) Weight (t = 1 min, 50 V) = 0.01 g, *Note:* Dye added to (uncoated) EAP to improve visualization. EAP surrounded by electrolyte solution. Electrode removed after experiment for weight determination [4]. Proc SPIE 2009, reprinted with permission

4.1 Ionic EAP Contraction

Selected ionic hydrogels composed of poly(methacrylic acid) (PMA) and other ion-containing polymers and copolymers produce good electroactive polymer (EAP) based materials and actuators that contract with low electric input (Fig. 4.1) [1–5]. A highly contractile EAP material was developed that is capable of pronounced contraction when subjected to an electric impulse, contracting over 80% (contracts to less than 20% of its original weight) in less than a minute at 50 V (Fig. 4.1). When the electricity is stopped, the flexible EAP relaxes back to its original size and shape [1–6]. In addition, some of these PMA based materials are able to quickly contract and then expand, repeatedly, by reversing the polarity of the electric input (Fig. 4.2) [3, 5]. By applying 50 V for 1 min and then reversing the polarity and applying the voltage for another minute, the EAP can contract and then expand back to and beyond its original size. This EAP contraction–expansion can be cycled (Figs. 4.2 and 4.3) [3, 5].

The mechanism of contraction is not well understood. The rapid weight loss and size reduction is believed to be primarily due to water leaving the EAP during contraction. Osada [7–27], Tanaka [28–40], Shahinpoor [41–52], Bar-Cohen [53–60], Khokhlov [61, 62], Shiga [63], and many others have investigated the behavior of hydrogels from various stimuli: change in pH, ionic concentration, change in solvent, and electric input. How and why this contraction phenomenon occurs in these contractile PMA based EAPs were determined using a radionuclide experiment, molecular modeling, and other experimental approaches.

In order for an ionic hydrogel-like material to be electroactive, several components are necessary:

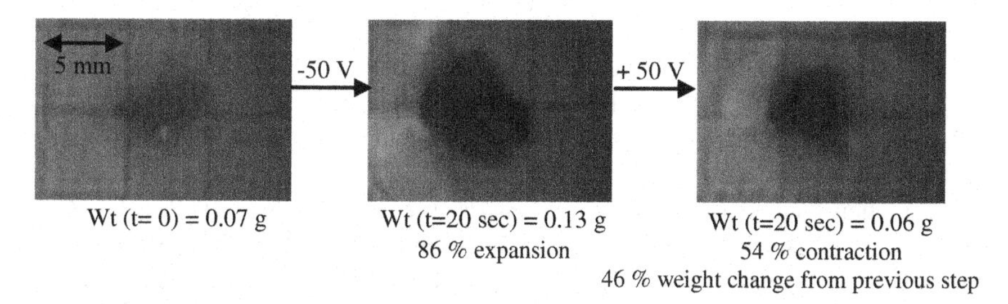

Fig. 4.2 Expansion-contraction cycle for a PMA based EAP. *Note:* Dye added to (uncoated) EAP to improve visualization. EAP surrounded by electrolyte solution [5], Proc SPIE 2011, reprinted with permission

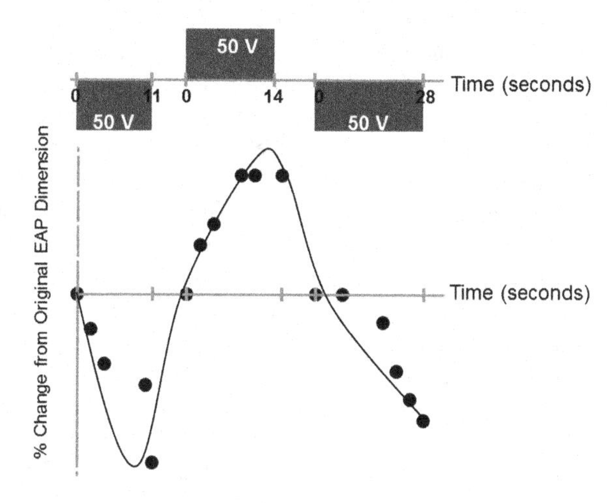

Fig. 4.3 Reverse polarity experiment [5], Proc SPIE 2011, reprinted with permission

1. The base polymer, or some component of the base polymer composition, must be ion-containing.
2. It is desirable for the EAP to be cross-linked or otherwise networked together.
3. The electrolytic solution that is absorbed by the EAP hydrogel plays a key role in electroactivity.
4. The placement and polarity of the electrodes helps dictate whether the EAP in the actuator bends, contracts, or expands.

There has been a great deal of interest worldwide in the search for materials that can transfer electrical energy directly into mechanical energy, analogous to our muscles converting electrical and chemical energy into mechanical energy to produce movement. For an ionic EAP, the polymers must be ion-containing, or for a copolymer, a portion of the copolymer must be ion-containing. Advances in these endeavors have been achieved using materials such as ionized poly(acrylamide), poly(acrylic acid), poly(acrylic acid)-co-(poly(acrylamide), poly(2-acrylamide-2-methyl-1-propane sulfonic acid), poly(acrylic acid), poly(methacrylic acid), poly(styrene sulfonic acid), quarternized poly(4.vinyl pyridinium chloride), poly(vinylbenzyltrimethyl ammonium chloride), and numerous other materials. Piezoelectric materials have also been investigated for use as electroactive materials; however, most piezoelectric materials undergo length changes of only a fraction of one percent.

Bar-Cohen's expertise with various EAPs is extensive [53–60], Pei [64–81], Smela [82–92], Anderson [93–104], De Rossi and Carpi [82, 105–116], Madden [117–119], and many others have developed and investigated electroactivity extensively in poly(pyrrole) and other dielectric based materials. SRI International and Artificial Muscle, Inc. hold many publications, [70–75, 78, 81, 120, 121] and patents, [122–126] in this area. Pelrine and Kornbluh at SRI International and Pei at UCLA have produced electrically driven mechano-chemical actuators, where the electric field is applied through flexible carbon plates, which provide for an expandable conducting surface, and an elastomeric material is sandwiched between the carbon plates. The elastomeric material wedged between the carbon plates acts as a flexible, movable structure when the two carbon plates, with opposing charges, are attracted and move closer to each other for the duration of the electric impulse. When the electric field is turned off, the material resumes its previous configuration. This type of electroactive actuation typically requires high voltages, and once the configuration is reached, the material is typically static. This strategy produces a very fast actuation.

Shahinpoor and Mojarred used ion-exchange materials and membranes to produce electrically responsive actuators [41–48] and also encapsulated ion-exchange membrane sensor/actuators. Shahinpoor used electrically responsive polymers coupled with springs and other mechanical devices to improve upon electrically responsive actuators [52] and ionic polymeric conductor composites and ionic polymer metal composites to drive pumps and mini-pumps [41].

Tanaka observed that ionized (polyacrylamide) (PAA) gels, immersed in a 50% acetone and 50% water mixture, collapsed and physically shrunk, by measured loss of volume, in the presence of an electric field. When the electric field was removed, the collapsed gel swelled to its initial proportions [28]. Tanaka observed phase transitions of cross-linked partially hydrolyzed PAA gels as a function of temperature, solvent, ionic concentration, pH, and electric field [28–40]. Very small changes in electric potential across partially hydrolyzed PAA gels produced significant volume changes [39, 40].

Osada and Hasebe found that water swollen poly(2-acrylamide-2-methyl-1-propane sulfonic acid) (PAMPS) gels contracted in the presence of an electric field, losing as much as 70% of their weight from loss of water [23]. Gels prepared from polymers and copolymers that contain ionizable groups, such as poly(methacrylic acid), partially hydrolyzed poly(acrylamide), poly(styrene sulfonic acid), quarternized poly(4-vinyl pyridinium chloride) and poly(vinylbenzyltrimethyl ammonium chloride), also exhibited marked contraction in the presence of an electric field, as did gels prepared from proteins such as gelatin and collagen, and gels prepared from polysaccharides, such as alginic acid and its salts, ager-ager and gum Arabic [7]. Osada observed that the presence of ionizable

Table 4.1 Comparison of hydrogel tensile strengths [6], Proc SPIE 2007, reprinted with permission

Material	Tensile strength (MPa)
Poly(acrylamide) gels	0.03
Poly(vinyl alcohol)-poly(acrylic acid) gels	0.23
Poly(2-hydroxyethyl methacrylate)-poly(methacrylic acid) crosslinked gels[a]	0.33

[a] 0.28–0.76 MPa range for these types of materials

groups in the polymer gels was a factor in predicting the response of a material to an electric field. According to Osada, polymer gels containing no ionizable moieties, such as poly(2-hydroxyethyl methacrylate) (PHEMA) and starch, showed no contraction in the presence of an electric field. [7, 9–12] Our work has also confirmed this, with no observed electroactivity in a variety of PHEMA formulations. Once enough PMA or other ion-containing polymers are introduced into a copolymerized formulation with PHEMA (or other neutrally charged polymeric component), and conditions are met in terms of required flexibility and other physical parameters, then electroactive PHEMA copolymers can be produced.

Using PAA or PAMPS, Osada developed mechanochemical actuators that "walked" in a looping fashion and a mechanochemical valve membrane that reversibly expanded and contracted its pore size when exposed to an electric field in alternate on/off cycles [20]. By doping cross-linked poly1N-13-(dimethyl-amino)propyl acrylamide with 7,7.8.8-tetracyanoquinodimethane, Osada produced materials that could undergo rapid contractions in the presence of an electric field [15, 19, 21].

Rasmussen found that copolymers comprising cross-linked networks of methacrylic acid and 2-hydroxy methacrylate, cross-linked with cross-linking agents such as ethylene glycol dimethacrylate and 1,1,1-trimethylolpropane trimethacrylate had tensile strengths well above the tensile strengths of polyacrylamide type materials (Table 4.1) [6].

Appendix A presents one of Prof. Osada's pivotal publications addressing a multitude of materials and their electroactivity, and just as important, classes of materials that are non-electrocative. Appendix B presents Prof. Tanaka's premier paper on the mathematical and physical theory of electroactivity based on ionized polyacrylamide gels.

For an ionic EAP, it is beneficial for the material to be crosslinked because linear ion-containing polymers simply dissolve into solution rather than retaining their shape and integrity. Crosslinking also improves the strength of these materials; however, this is a delicate balancing act. If the hydrogel is too crosslinked, it may be very strong, brittle even, but it loses flexibility if too highly crosslinked. In order to be able to undergo movement when electrically stimulated, the EAP must be flexible. Conversely, if the material is too lightly crosslinked, the EAP, though very flexible, will be inherently too weak and will undergo extremely substantial swelling in the presence of solvent, further weakening the material. When cross-linking these EAPS, there is a trade-off between strength and flexibility. Different

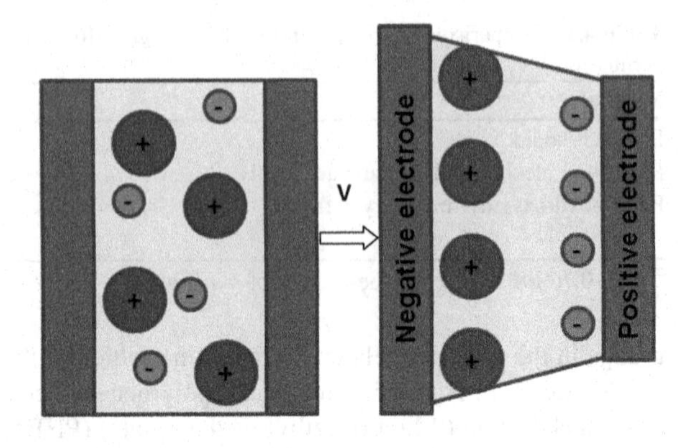

Fig. 4.4 Traditional ionic EAP configuration

levels of crosslinking can be used to tailor the amount of electroactivity vs. inherent strength of the EAP for different end use applications. For a non-load bearing application, such as a valve, a highly contractile EAP may be desirable because strength is not as crucial. For a load bearing application, such as an EAP based actuator in a robotic or prosthetic device, the electroactivity may need to be sacrificed slightly in order to acquire the necessary high strength needed to make the EAP practical in these types of real world applications. Interpenetrating network technology also provides for a way to couple electroactivity and strength within the same EAP based actuator.

Most EAPs bend in an electric field. The greater the size differential between the negative and positive ions of the salt in the electrolyte solution contained in the EAP, the more pronounced the bending (Fig. 4.4). Alternatively, if one of the ion types is very large and relatively immobile, while the opposite charged ion type is very small and mobile, bending can also occur in an electric field from the mobility of the smaller mobile ions. Reversing the polarity causes an oscillatory motion, similar to the back-and-forth motion of a fish tail propelling the fish forward through water.

So why do some ionic EAPs contract? The movement of the ions of the electrolyte comes into play, but to exploit contraction, the placement of the electrodes is also very important. If contraction, rather than bending is desired, the negative electrode is placed external to the EAP, while the positive electrode remains internal (embedded) in the EAP (Fig. 4.5). In the case of the contractile EAP composed of PMA, the ions along the main polymer chains are weak acetic acid groups. The electrolytic solution typically used to swell the gel and provide for good electroactivity is a dilute sodium chloride solution, such as isotonic saline solution. When the electric input is applied to the PMA EAP based actuator with this electrode configuration, the EAP of the actuator contracts quickly and substantially. In selected EAP formulations, when the polarity is reversed, expansion occurs, and contraction and expansion can be cycled repeatedly in selected PMA based EAP materials and actuators [3, 5].

A radionuclide experiment was performed to determine, in as close to real time as possible, the chain of events that occurs when electricity is applied to EAP

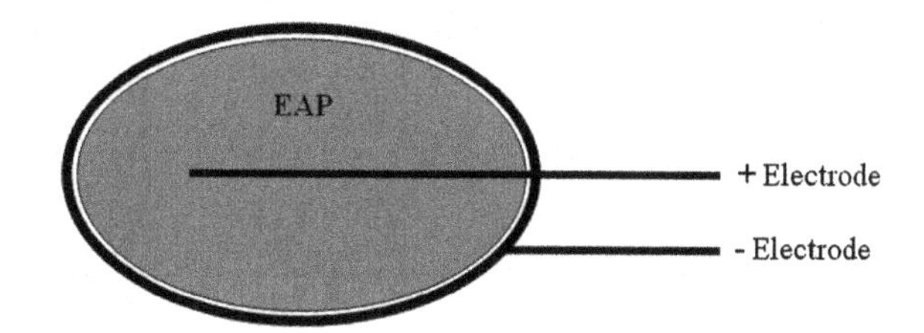

Fig. 4.5 Electrode configuration for contractile EAP

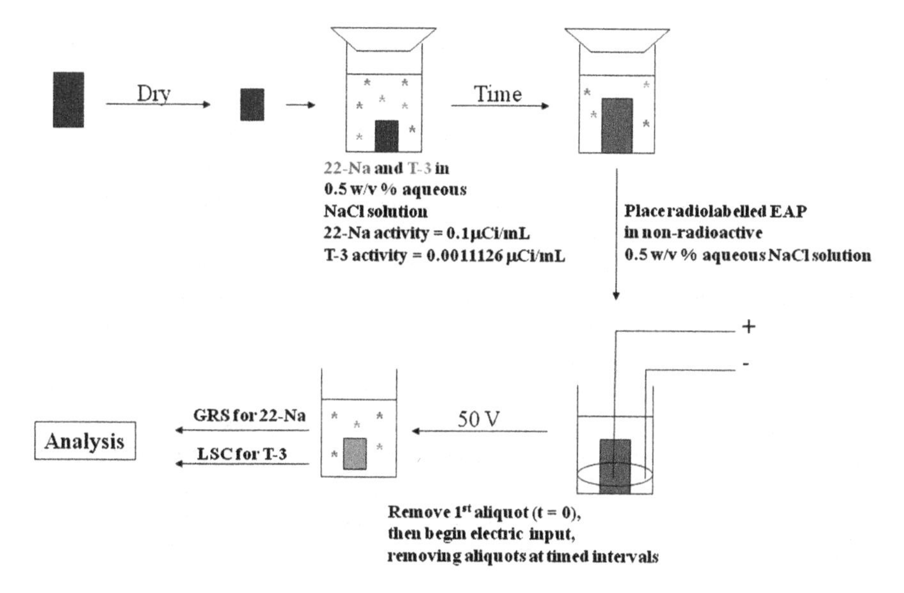

Fig. 4.6 Radionuclide experiment to follow EAP contraction [2], Proc SPIE 2010, reprinted with permission

materials, both in terms of water loss and electrolyte flow during contraction (Fig. 4.6). EAP samples were immersed in aqueous NaCl solutions containing known concentrations for both tritium (H-3) and sodium-22 (Na-22) radionuclides and then allowed to swell and equilibrate. For the electroactive experiment, EAP samples were then placed in an unlabeled aqueous NaCl solution and subjected to electric input. Aliquots of the surrounding solution were removed in timed intervals during the electric input. Control experiments were also performed, where no electricity was applied to sample EAPs while aliquots were removed in the same timed intervals. Liquid scintillation chromatography (LSC) was used to detect tritium levels and gamma ray spectroscopy (GRS) was used to detect Na-22 levels.

The results from the radionuclide experiment indicate that during the electric input, Na-22 left the contractile EAP very quickly, with significant outflow of the positive sodium ions towards the negative electrode within 5 s. The Na-22 outflow plateaued around 75 s. A similar profile was seen with tritium. Based on the control

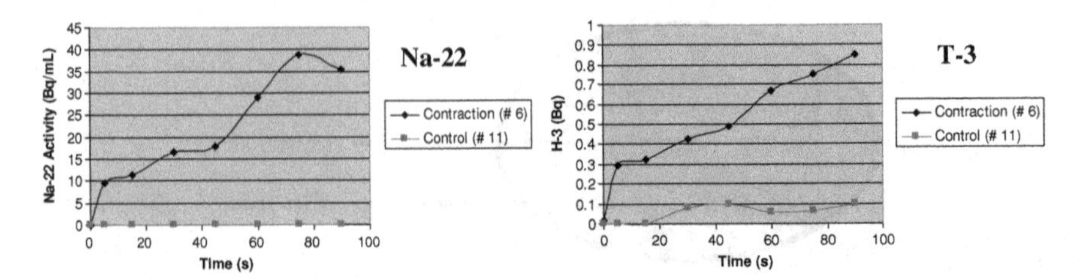

Fig. 4.7 Radionuclide-labeled contraction experiments following Na-22 and T-3 [2], Proc SPIE 2010, reprinted with permission

(no electric input), the Na-22 was retained very tightly to the EAP material, while the tritium had some baseline mobility in and out of the EAP. Once the EAP was activated by electricity, the positive sodium ions and water moved very quickly and simultaneously out of the EAP material, leading to a very fast, pronounced contraction (Figs. 4.1 and 4.7). Indeed, it appears that electricity is required in order for sodium ions to flow out of these contractile EAPs. Following water loss by volume change and weight conversions in real time showed a similar trend (Fig. 4.8).

After the electric input part of the radionuclide-labeled experiment, the EAPs were placed in fresh unlabeled aqueous sodium solution, allowed to equilibrate for several days, and then aliquots taken from each equilibration media (Fig. 4.9). GRS found no Na-22 activity in the equilibration media, while tritium levels were fairly high, particularly for the control (Table 4.2). Again, this indicates that the Na-22 is very tightly bound to the contractile EAPs when there is no electricity.

Finally, the EAPs themselves were tested using GRS to account for the location for all Na-22, particularly in the controls which had no electric input. After both the initial contraction experiment and the equilibration experiment, the Na-22 that was not released during the previous experimentation remained in the EAP gels, with 0.6425 Bq remaining in the EAP that had undergone contraction and 3.834 Bq remaining in the control EAP (Fig. 4.10).

Based on the radionuclide experiment, the sodium ions do not move at all unless electrically stimulated. The net outflux of sodium ions to the externally placed negative electrode also causes a huge outflux of water molecules, virtually simultaneously, causing a rapid and profound contraction of the EAP. In selected formulations, when the polarity is reversed, rapid expansion of the EAP occurs, and the contraction–expansion can be cycled repeatedly.

So the distinct movement of sodium ions and water molecules that occurs during electric impulse causes EAP contraction, but why is this cation migration, and subsequent water molecule migration, occurring? Molecular modeling excluded the hypothesis of selective mobility due to pore sizes within the EAP. This is because in order to be flexible enough to be electroactive, the pore sizes in these cross-linked PMA based EAPs are huge compared to any of the electrolyte systems investigated, including ionic liquids (Fig. 4.11).

Different ions move at different rates within an aqueous environment. For simple electrolytes composed of a Group 1 cation associated with a group 7 anion,

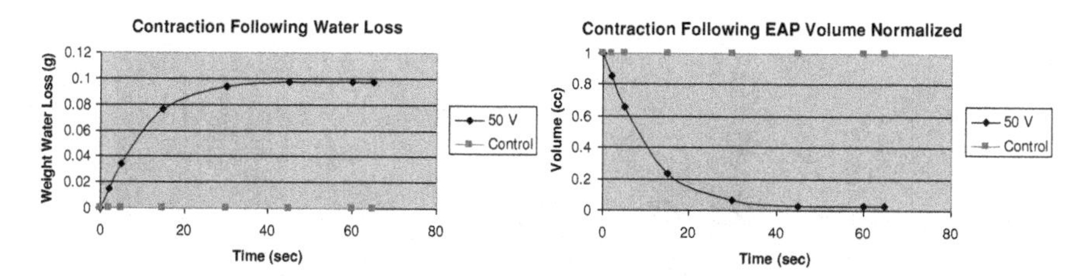

Fig. 4.8 Contraction following water loss from weight change and from volume change [2], Proc SPIE 2010, reprinted with permission

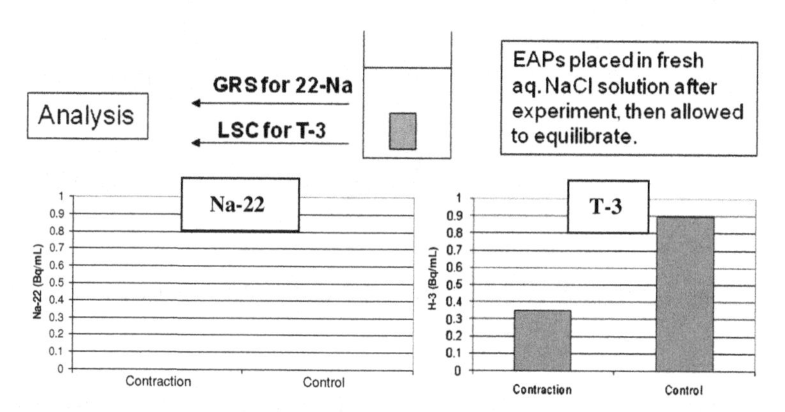

Fig. 4.9 Radionuclide equilibration following contraction [2], Proc SPIE 2010, reprinted with permission

Table 4.2 Radionuclide-labeled equilibration results following contraction experiment

Sample	Na-22 (Bq/mL)	T-3 (Bq/mL)	Normalized T-3[a]
Control	0	0.8933	100
Contraction EAP	0	0.3475	34

[a] Normalized with respect to the control and to the initial weight of the EAPs

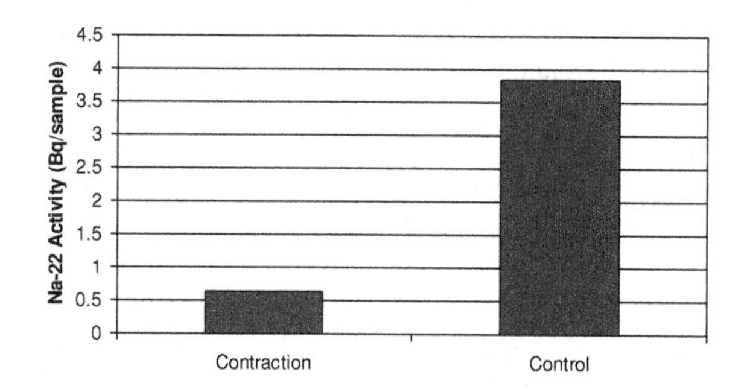

Fig. 4.10 GRS analysis of the EAPs themselves following Na-22 [2], Proc SPIE 2010, reprinted with permission

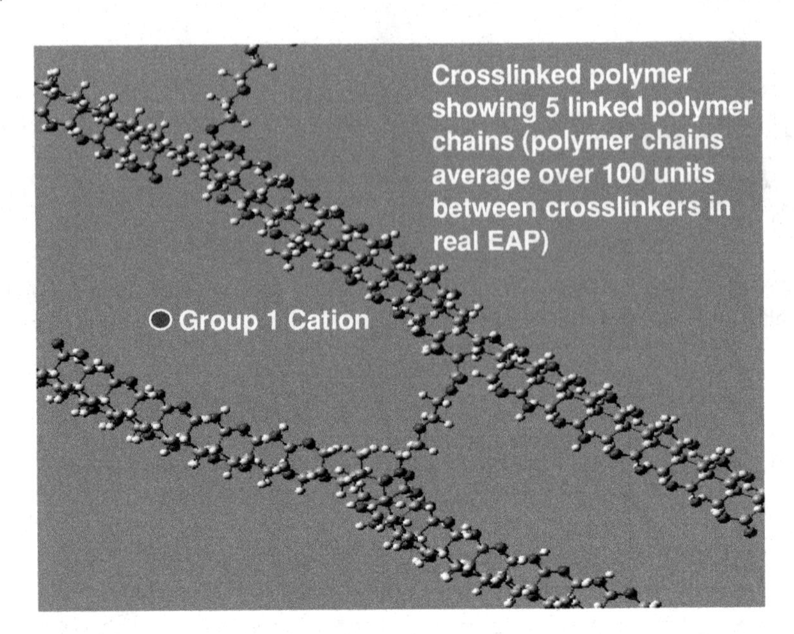

Fig. 4.11 Molecular model of PMA based EAP [5], Proc SPIE 2011, reprinted with permission

such as LiCl, NaCl, KCl, LiBr, NaBr, and KBr, the cation moves much faster in an electric field than its associated anion. The movement of cations and the subsequent movement of water was behind the profound levels of contraction observed in these hydrogels.

Based on the sizes of the Group 1 ions lithium, sodium, and potassium, one may predict that the lithium ion, with its tiny atomic radius of 0.69 Å, would migrate the fastest in an electric field, thus producing the highest level of contraction in an EAP. However, in an aqueous environment, lithium has the highest hydrodynamic radius (2.38 Å) compared to the other Group 1 cations, so in fact moves the slowest in an electric field (Fig. 4.12 and Table 4.3) [127–130] and produces the least amount of contraction (Fig. 4.13). Conversely, the potassium ion, with its large atomic radius of 1.25 Å but small hydrodynamic radius of 1.25 Å (virtually same as atomic radius) moves very quickly in an electric field (Table 4.3) [127–130] and produces the highest amount of contraction (Fig. 4.13).

Experimentally, three PMA based EAPs (from the same batch) were swollen in three different Group 1-Group 7 dilute electrolyte solutions: 0.2 M LiBr, 0.2 M NaCl, and 0.2 M KCl. The EAPs were then subjected to a 50 V, 200 mA electric input for 3 min, with the positive electrode embedded in the EAP and the negative electrode external to the EAP. The EAPs swollen with potassium chloride produced the fastest and most pronounced contraction of 32%, followed by sodium chloride at 26% contraction, and followed by lithium bromide at 21% contraction (Fig. 4.13).

These were very dilute electrolyte solutions. As the electrolyte concentration is raised, the electroactivity improves, up to a point. For methacrylic and acrylic

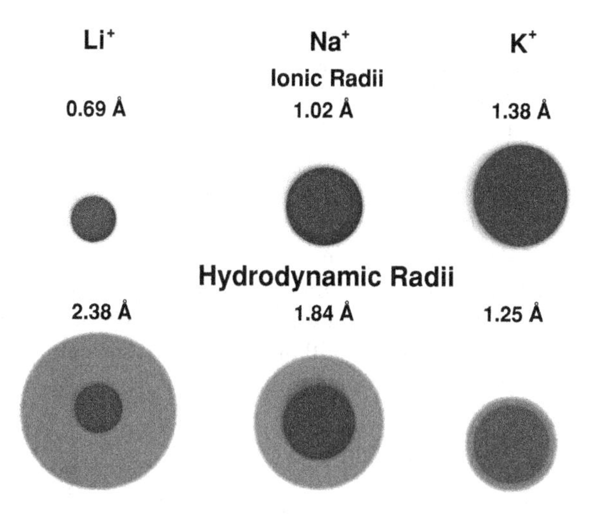

Fig. 4.12 Ionic and hydrodynamic radii for lithium, sodium and potassium Group 1 cations

Table 4.3 Ionic crystallographic radii, hydrodynamic radii, molar equivalent conductivities, and diffusion coefficients of lithium, sodium, and potassium cations at ∞ dilution at 25°C [127–130]

Group 1 (Alkali) Metal	Ionic radii r_c (Å)	Hydrodynamic radii r_h (Å)	Molar conductivities $\lambda^{\infty} \times 10^4$ (m^2S/mol)	Diffusion coefficients $D \times 10^9$ (m^2/s)
Li^+	0.69	2.38	38.66	1.029
Na^+	1.02	1.84	50.08	1.334
K^+	1.38	1.25	73.48	1.957

Fig. 4.13 Contraction of PMA based EAPs for three different electrolyte systems [5], Proc SPIE 2011, reprinted with permission

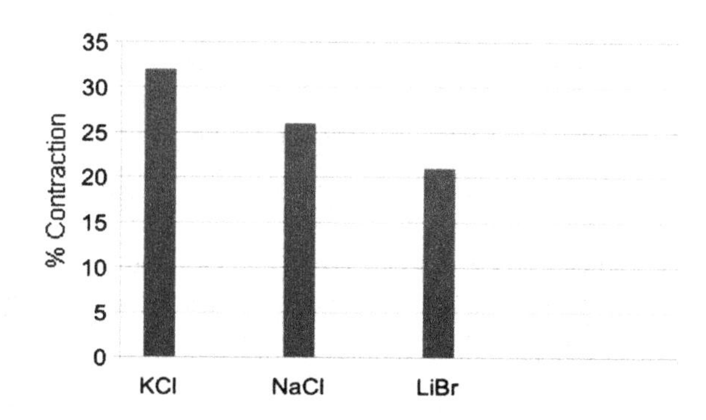

based EAPs, if the electrolyte concentration is too high, the polymer undergoes a phase change, the polymer morphology changes, becoming stiff (chalk like texture), and then is much less or even simply not electroactive due to the loss in flexibility. Dilute electrolyte solutions work well.

Besides the cations being attracted to the very negative external electrode during electric impulse, there is also a local electrochemistry effect at each electrode, which causes chemical changes in these PMA based EAPs. In an aqueous environment, NaCl dissociates into sodium and chloride ions:

$$NaCl \rightarrow Na^+ + Cl^-$$

At the anode, which is the embedded positive electrode, two things are occurring. Some of the chloride anions in this dilute solution are becoming oxidized and forming a small amount of chlorine, while water is oxidized and forming oxygen and protons at the embedded positive electrode. The formation of protons at the positive embedded electrode within the EAP dramatically lowers the pH. This produces a very acidic environment for the EAP.

$$2\,Cl^- \rightarrow Cl_2 + 2\,e^-$$

$$2\,H_2O \rightarrow O_2 + 4\,H^+ + 4\,e^- \quad \text{Acidic at Anode}$$

Meanwhile at the cathode, which is the external negative electrode, water undergoes reduction, releasing hydrogen at the cathode and forming hydroxide ions, which raises the pH, producing a basic environment.

$$2\,H_2O + 2\,e^- \rightarrow H_2 + 2\,OH^- \quad \text{Basic at Cathode}$$

How does the localized electrochemistry affect these EAPs? Recall that these PMA based EAPs contain weak acid groups along the main polymer chains (Figs. 4.11 and 4.14). In an acidic environment, the acid groups are neutralized by the input of protons during the electric impulse on the anode, which is the embedded positive electrode, so the EAP becomes neutrally charged (Fig. 4.14). The positively charged cations, such as sodium ions, leave the EAP from this two-fold effect:

1. The cations are very attracted to the cathode, which is the external negative electrode.
2. The cationic counter ions now have no ionic affiliation with the protonated acid groups of the EAP, because the acidic groups of the EAP lose their charge and become neutral during electric impulse at the anode, which is the embedded internal anode.

As the cations migrate rapidly out of the EAP, the water molecules in the hydration layers around the sodium ions are also dragged away from the EAP, resulting in contraction.

When the polarity is reversed, the strong negative charge is now at the embedded electrode and the localized electrochemistry is producing a basic environment within the EAP, thus the EAP regains its negative charges along the polymer chains. The cations, such as sodium ions, return to the EAP when the polarity is reversed, again from this two-fold effect:

In an aqueous solution, NaCl disassociates into Na^+ and Cl^- ions
$NaCl \rightarrow Na^+ + Cl^-$
At anode (positive electrode)
$2\,Cl^- \rightarrow Cl_2(g) + 2\,e^-$
As to water itself, at anode
$2\,H_2O \rightarrow O_2(g) + 4\,H^+ + 4\,e^-$ **Acidic at anode**

At cathode (negative electrode)
$2\,H_2O + 2\,e^- \rightarrow H_2(g) + 2\,OH^-$
$2\,OH^- + 2\,Na^+ \rightarrow 2\,NaOH$ **Basic at cathode**
As to water itself, at cathode
$4\,H+ + 4\,e- \rightarrow 2\,H_2(g)$

Overall electrolysis reaction for NaCl dissolved in H_2O
$2\,NaCl + 2\,H_2O \rightarrow Cl_2(g) + H_2(g) + 2\,NaOH$ $E_o = -1.36$ V
Overall electrolysis reaction for H_2O
$2\,H_2O \rightarrow 2\,H_2(g) + O_2(g)$ $E_o = -1.23$ V

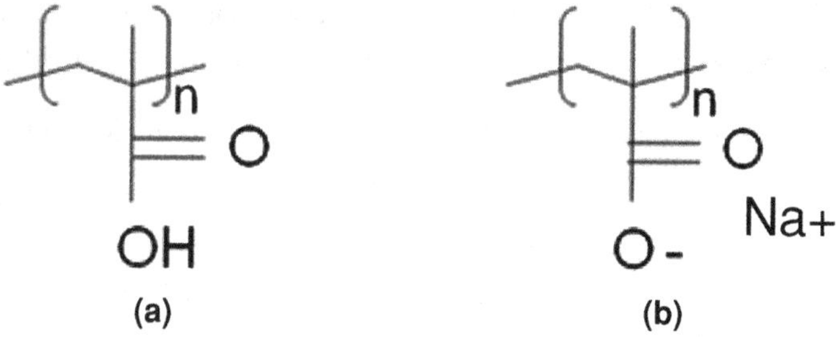

Fig. 4.14 Local electrochemistry at the electrodes. **a** Dominates in an acidic environment **b** Dominates in a basic environment

1. The migration of the positively charged cations to the (embedded) negative electrode and away from the (external) positive electrode.
2. The local electrochemistry within the EAP returns the negative charge to the weak acid groups along the polymer chains as the acid groups are deprotonated in the now basic local environment, which also causes cationic migration back into the EAP.

As the cations migrate rapidly back into the EAP, the water molecules in the hydration layers around the cations are simultaneously dragged back into the EAP, resulting in expansion.

The electrolyte concentration in these EAPs is very low, though, so movement purely by the ions and associated water shells does not fully explain the pronounced contraction observed in these EAPs when electrically stimulated. In fact, if the water shell migration were the only water movement occurring, lithium

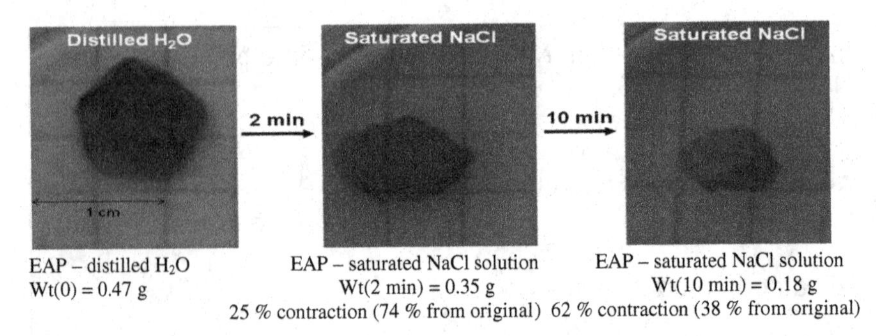

EAP – distilled H_2O
Wt(0) = 0.47 g

EAP – saturated NaCl solution
Wt(2 min) = 0.35 g
25 % contraction (74 % from original)

EAP – saturated NaCl solution
Wt(10 min) = 0.18 g
62 % contraction (38 % from original)

Fig. 4.15 Osmotic effects of PMA based EAP [5], Proc SPIE 2011, reprinted with permission

would be expected to have the highest contraction of the Group 1 cations due to the large amount of hydration layers (Fig. 4.12). The converse is observed, where potassium, in fact, has the highest level of contraction compared to sodium and lithium cations. The contraction observed once ion migration is underway can be explained, however, by osmosis.

Once the sodium ions migrate out of the EAP in response to the electric impulse, then an ionic concentration gradient occurs. Due to osmosis, water rapidly leaves these flexible EAPs due to the higher ion concentration outside of the EAP once the sodium ions have migrated external to the EAP due to the placement of the cathode (the external negative electrode) and due to the local pH effects at the anode (the embedded positive electrode).

The osmotic effect was experimentally determined by placing the EAP first in a distilled water environment, followed by placing the EAP into a saturated sodium chloride solution and then observing any changes in the EAP. When the EAP was placed into the saturated NaCl solution, within 2 min the EAP had contracted 25%. Within 10 min, the EAP had contracted by 62% (Fig. 4.15).

The contraction that the PMA based EAPs undergo when electrically stimulated is thus a three-fold effect due to the following:

1. The strong negative charged cathode, which is placed external to the EAP, attracts the positively charged cations, such as sodium ions, out of the EAP.
2. The localized pH effects at the positive charged anode, which is embedded in the EAP, protonates the weak acid groups along the polymer chains of the EAP, making the EAP neutrally charged, so that the cations are no longer attracted to the side groups of the EAP.
3. The osmotic effects following the ion gradient from the rapid cation migration in an electric field cause a large water loss out of the flexible EAP.

All three of these effects cause a rapid and pronounced contraction in selected EAPs. The first two events occur in tandem, followed quickly by the third effect of osmosis, causing a rapid and pronounced contraction in these PMA based EAPs.

Reversing the polarity causes expansion by a similar three-fold effect due to the following:

1. The strong negative charged cathode, which is now the internal embedded electrode in the EAP, attracts the positively charged cations, such as sodium ions, back into the EAP.
2. The localized pH effects at the negatively charged cathode, which is now embedded in the EAP, deprotonates the weak acid groups along the polymer chains of the EAP, making the EAP negatively charged again, so the cations are attracted back to the side groups of the EAP.
3. The osmotic effects following the ion gradient from the rapid cation migration in an electric field cause a large water gain back into the flexible EAP.

4.2 Extreme Temperature Considerations

These PMA based contractile EAPs were able to withstand extreme temperatures. Cryogenic experiments were performed by subjecting the contractile EAPs to 4.2 K using liquid helium, 77.25 K using liquid nitrogen, 194.65 K using a dry ice/isopropyl alcohol bath, and 273.15 K using an ice bath. Elevated temperature experiments were conducted up to 410 K. Exposing these contractile EAPs to extreme temperature conditions, even extremely cold temperatures, did not affect their ability to contract when electrically stimulated and had no discernable adverse effects. In the case of extreme cold exposure, these EAP actually performed better once thawed [4]. In order to actuate under extreme cold temperatures, the electricity can be ramped up to provide heat for a quick thaw, and then ramped down to actuation levels.

4.3 Carbon Fiber Infused EAPS

Besides plasma treated metal electrodes, carbon fibers also adhere well to these PMA based EAPs. PMA based EAPs with embedded carbon fibers, arranged in long linear patterns or in grid patterns, respond very well to electricity, with marked contraction at 50 V with over 50% contraction by weight loss in a minute or less. Small cylindrical EAPs with embedded carbon fibers (Fig. 4.16) contracted and expanded when the polarity was reversed, albeit slowly, at voltages even as low as 1 V.

4.4 Fourth Generation Contractile EAPs

The presence of Group 7 electrolytic anions is disadvantageous due to their slow migration and electrophoretic affect at the anode. Contraction-expansion cycles for a robust PMA based copolymer, which only contains cations affiliated with the acidic groups along the main EAP chain (no Group 7 or other free anions are present) is shown in (Fig. 4.17). Besides electrolytic solution, this copolymer has

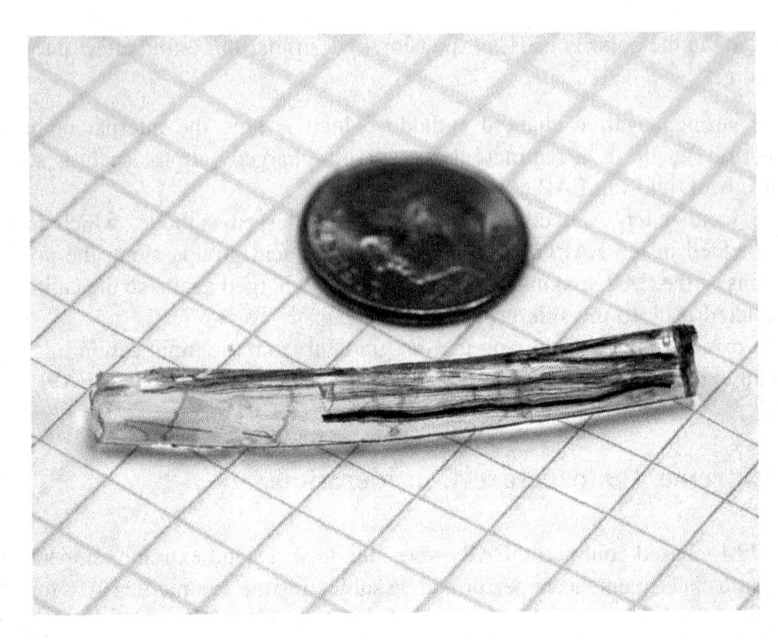

Fig. 4.16 Small cylindrical EAP with embedded carbon fibers

Fig. 4.17 Fourth generation EAP copolymer in water

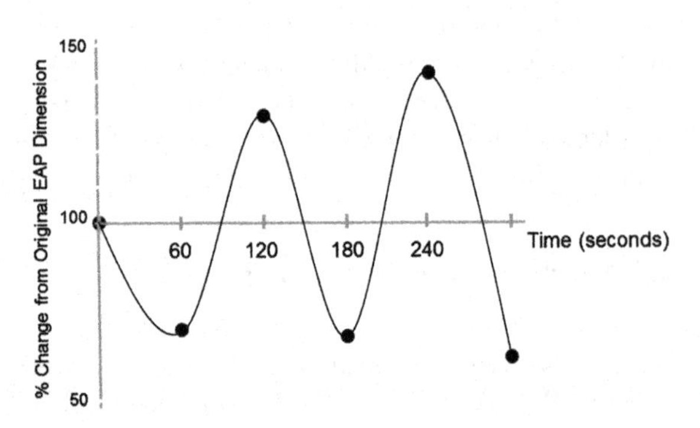

an advantage of operating extremely well in plain tap water (or in distilled water with a minute amount of electrolyte added to provide for conductivity).

4.5 Overview Ionic EAPs Capable of Contraction and Expansion–Contraction Cycles

Robust ionic electroactive materials have been synthesized, where the strength and electroactivity can be tailored by controlling the extent of crosslinking and other synthetic strategies. These EAP based materials and actuators can undergo rapid and pronounced contraction. Reversing the polarity of selected EAPs produces contraction–expansion cycles. Both the highly contractile EAP and the EAP capable of fast contraction–expansion cycles lend themselves to smooth, controlled, three-dimensional, life-like, biomimetic motion.

Radionuclide-labeled electrolyte (Na-22) and tritium (T-3) labeled water were used to determine the chain of events that occur when these PMA based EAPs contract. With electric input, cations, such as sodium ions, are attracted to the external negative electrode, while local electrochemistry and pH changes affect these EAPs at the embedded positive electrode. The osmotic effects following the ion gradient from the rapid cation migration in an electric field cause a large water loss out of the flexible EAP. The cations and water leave the EAP, causing a rapid collapse and contraction of the EAP. Without electric input, the cations remain very tightly bound to the ionic EAPs. Once electricity is applied to these EAP based materials and actuators, cations and water move very rapidly from a threefold effect due to ion migration, localized pH effects, and osmosis, producing a pronounced contraction of the EAP. Reversing the electric polarity of the external and internal electrodes causes a reversal in the chain of events, producing a significant expansion of the EAP.

4.6 Practical Considerations when Designing Actuators Using Ionic EAPs

Once a very electroactive material was developed, the next challenge was how to keep an actuator electronically wired using these types of dynamic materials. Keeping the electrodes placed in the EAP during movement is an extreme challenge because when the EAPs undergo motion the electrodes that are attached to them, even if embedded, can become detached from the EAPs, which causes actuator failure. Plasma treatment and other metal treatments were investigated to improve the interface between the EAP and the embedded electrodes.

Plasma is partially or wholly ionized gas with about an equal number of positively and negatively charged particles. Some scientists have named plasma the "fourth state of matter." While plasma is neither gas nor liquid, the properties of plasma are similar to those of both gases and liquids. Sterilization and improving the adhesion between two surfaces are common applications. Plasma surface treatment can create chemically active functional groups such as amine, carbonyl, hydroxyl and carboxyl groups, which can greatly improve interfacial adhesion. Plasma is used to improve bonding on substrates such as glass, polymers, ceramics and metals. By improving the metal–polymer interface, the EAP material and electrode(s) in an actuator can work as a unit, where the electrode(s) delivers the electric input, much like a nerve, and can also serve as a tendon between the EAP material and a lever.

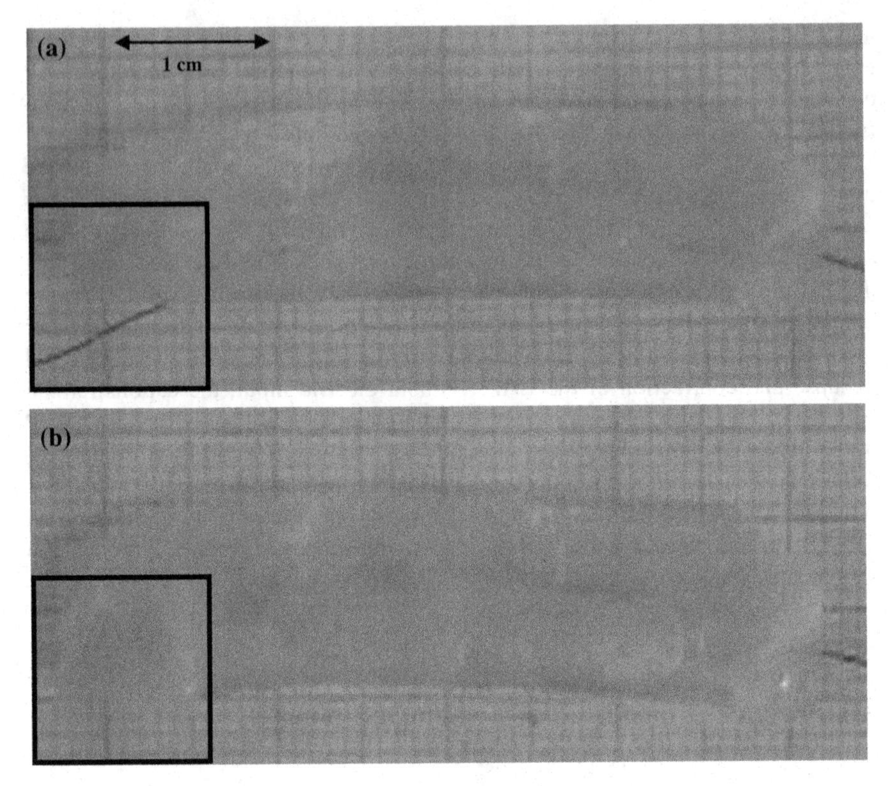

Fig. 4.18 Coated EAP material with embedded electrodes [6], Proc SPIE 2007, reprinted with permission

For actuators designed to act like an artificial muscle, EAPs were synthesized with at least one embedded electrode. If the material is particularly electroactive, the material would move so quickly that the embedded wire would disengage, causing the actuator to lose its electric impulse (Figs. 4.18 and 4.19).

Plasma treatment was investigated to improve the interface between the electrode metal and the electroactive material. Plasma can be generated in a near vacuum to form a variety of reactive species—positive and negative ions, excited-state species and radicals—as well as photons and neutral species from recombination. Aircraft-grade 302/304 stainless steel and Grade 2 titanium were subjected to plasma treatment. Typically, plasma was generated by a 25 MHz RF power source operating between 16 and 20 W. Gas was streamed in between 2 and 5 mTorr pressure and plasma maintained for several hours. Samples were clamped to a long I-probe, which allowed for the samples to be connected to a negative potential and placed mid-stream in the plasma pathway of a plasma chamber. A negative potential preferentially attracted positive ions to the sample surfaces. Several testing methods and surface analyses were used to characterize the metallic surface and determine the strength of the metal–polymer interface.

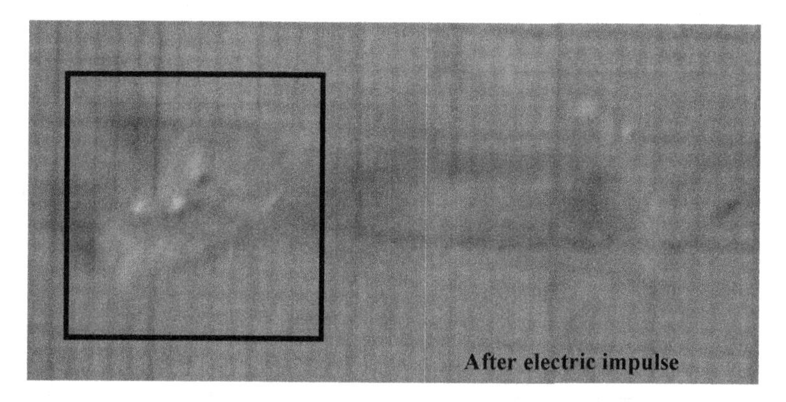

Fig. 4.19 Damage to electroresponsive material in boxed area. *Note:* Coating removed to aid visualization [6], Proc SPIE 2007, reprinted with permission

The water drop surface contact angle test uses a (distilled) water drop on a surface and measures the contact angle. The monomer mixture for the EAP is very hydrophilic. Surfaces with a low contact angle using a water drop are desirable for good metal-polymer cover and adhesion because the monomer mixture then glides uniformly across the metal surface, rather than beading up. The untreated stainless steel had an average water drop surface angle of 81°. Helium and hydrogen plasma treatment did not improve the water drop surface contact angle (117 and 96°, respectively). Nitrogen and synthetic air plasma treatment markedly improved the water drop contact surface angle (26 and 40°, respectively). Using pure oxygen plasma, however, produced the most hydrophilic stainless steel surface (4° contact angle; Table 4.4). The untreated titanium had an average water drop surface angle of 81°. Helium, nitrogen and synthetic air plasma treatment markedly improved the water drop contact surface angle (55, 47, and 23°, respectively). Just like stainless steel, using pure oxygen plasma to treat titanium produced the most hydrophilic titanium surface (5° contact angle; Table 4.4).

X-ray photoelectron microscopy (XPS) was used to determine the atomic surface composition of plasma-treated stainless steel foil by analyzing binding energies for Fe 2p, O 1, N 1 and C 1 s orbitals (Fig. 4.20 and Table 4.5). For carbon, both the nitrogen and oxygen plasma treatments served to significantly reduce the presence of carbon (C 1 s C–C peak at 285.3 eV). The presence of carbon is indicative of oils and other organic contaminants. Plasma can strip, sterilize, and super-clean metallic surfaces. Nitrogen is difficult to detect using XPS. For nitrogen, there was no discernible peak in the control sample. After nitrogen plasma treatment, a weak peak 396.2 eV (N 1 sN–Fe peak) was detected. This N–Fe peak was reduced after the oxygen plasma treatment. For oxygen, the oxygen peaks were bimodal, roughly split between the hydroxyl O–H peak (O 1 s peak at 533.1 eV) and the iron oxide peak (O1speak at 530.5 eV; Fig. 4.21). After nitrogen plasma treatment, the O1 s peaks were markedly elevated, particularly in the oxide region, with a shoulder extending into the hydroxyl region.

Table 4.4 Water drop contact angle test on plasma-treated metal surfaces [4], Proc SPIE 2009, reprinted with permission

Treatment	Stainless steel ()	Titanium	Titanium
Control	81	81	
N plasma	26	47	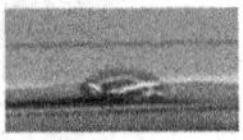
O plasma	4	5	

Again, this was indicative of the nitrogen plasma stripping the metal surface, exposing the iron oxide layer, which was also evident from the C 1 s results. After subsequent oxygen plasma treatment, the O 1 s peaks, which were bimodal, were markedly elevated, particularly in the hydroxyl region, indicating an oxygen-based chemical modification at the atomic level on the plasma-treated stainless steel surface. The high level of hydroxyl groups in addition to a good iron oxide surface provided for a very hydrophilic metal surface, which was also evident from the water drop contact angle tests. For iron, the nitrogen plasma-treated sample had the highest peak (Fe 2p1/2 at 725 eV and Fe 2p3/2 at 711.5 eV, both peaks from iron oxides). Again, this was indicative of the nitrogen plasma stripping the metal surface, exposing the iron oxide layer, which was also evident from the C 1 s and O 1 s results. After oxygen plasma treatment, the iron oxide peaks were slightly diminished, probably due to the pronounced oxygen layer from hydroxyl groups. The nitrogen plasma chemically modified the stainless steel surface somewhat, but more importantly, was paramount in cleaning the metal surface of contaminants. Argon could also be used for this purpose. The subsequent oxygen plasma chemically modified the stainless steel surface with oxygen groups in addition to providing for a good, clean iron oxide surface.

Titanium was analyzed with XPS to determine the atomic surface chemistry of plasma treated titanium foil by analyzing binding energies for O 1 s, Ti 2p, N 1 s and C 1 s orbitals (Fig. 4.22 and Table 4.5). The XPS results for titanium were similar to the results found for stainless steel. For titanium, both the nitrogen and oxygen plasma treatments served to significantly reduce the presence of carbon (C 1 s C–C peak at 285.3 eV, in control sample, small C–Ti peak at 282.2 eV). After oxygen plasma treatment, in addition to a small Ti–N peak at 397.3 eV, there was an additional small, broad peak centered around 400.5 eV, which is indicative of an N–O bond (TiO_2 peak). For oxygen, the oxygen peaks were bimodal, roughly split between the hydroxyl O–H peak (O 1 s peak at 533.1 eV) and the titanium oxide peak (O 1 s peak at 530.8 eV; Fig. 4.23). After nitrogen plasma

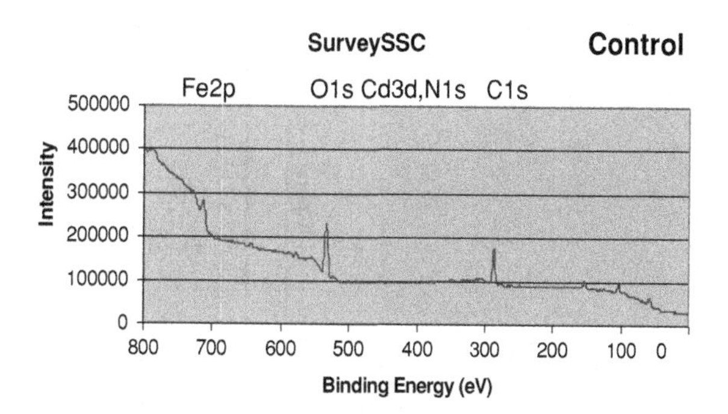

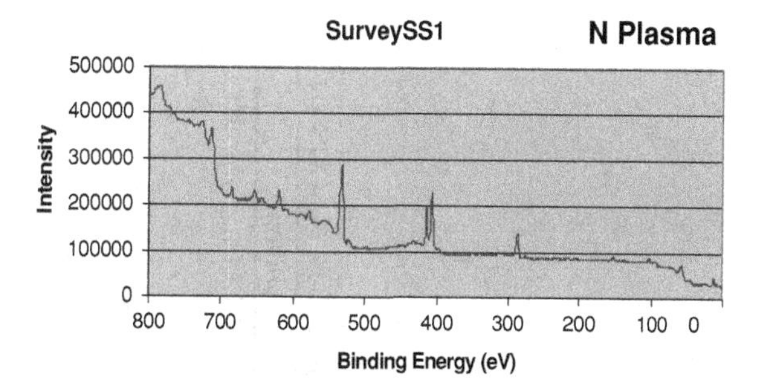

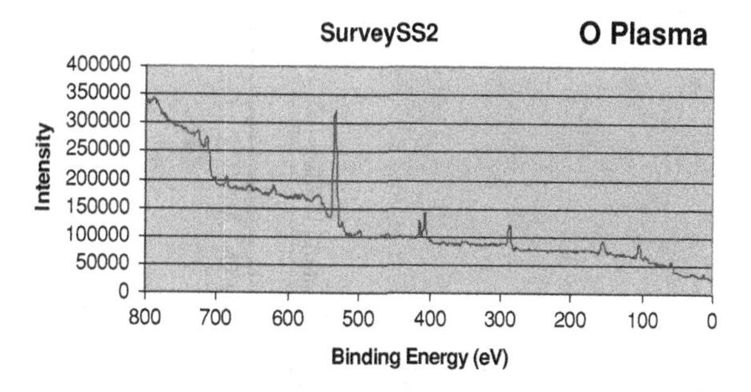

Fig. 4.20 Stainless steel XPS spectra [4], Proc SPIE 2009, reprinted with permission

treatment, the O 1 s peaks were markedly elevated, particularly in the oxide region, with a shoulder extending into the hydroxyl region, which was indicative of the nitrogen plasma stripping the metal surface, exposing the titanium oxide layer. After subsequent oxygen plasma treatment, the O 1 s peaks were markedly elevated, particularly in the hydroxyl region, with a shoulder extending into the oxide region, indicating an oxygen-based chemical modification at the atomic level on the plasma treated titanium surface. The high level of hydroxyl groups in addition to a titanium oxide surface provided for a very hydrophilic metal surface, which, like stainless steel, was also evident from the water drop contact angle tests. For titanium, the nitrogen plasma treated sample had the highest peaks (Ti 2p1/2 at 465 eV and Ti 2p3/2 at 459.2 eV, both peaks from TiO_2). Again, this was indicative of the nitrogen plasma stripping the metal surface, exposing the titanium

Table 4.5 Relative atomic composition of plasma treated stainless steel and titanium surfaces from XPS [1], Polym Int, reprinted with permission

	Stainless Steel					**Titanium**				
Plasma treatment	C1 s $C_{C\text{-}C,\ C\text{–}H}$	N1 s $N_{Fe\text{-}N}$	O1 s $O_{Fe\text{-}OH}$	O1 s $O_{Fe\text{-}O}$	Fe2p Fe	C1 s $C_{C\text{-}C,\ C\text{–}H}$	N1 s $N_{Ti\text{-}N}$	O1 s $O_{Ti\text{-}OH}$	O1 s $O_{Ti\text{-}O}$	Ti2p Ti
Control	**45**	–	**22**	**24**	**24**	**38**	**68**	**20**	**28**	**23**
Nitrogen	**29**	**76**	**26**	**43**	**50**	**32**	**21**	**34**	**58**	**66**
Oxygen	**26**	**24**	**52**	**33**	**26**	**30**	**11**	**46**	**14**	**11**

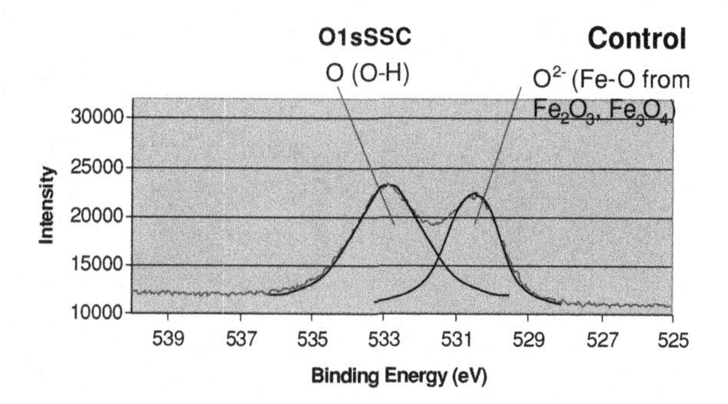

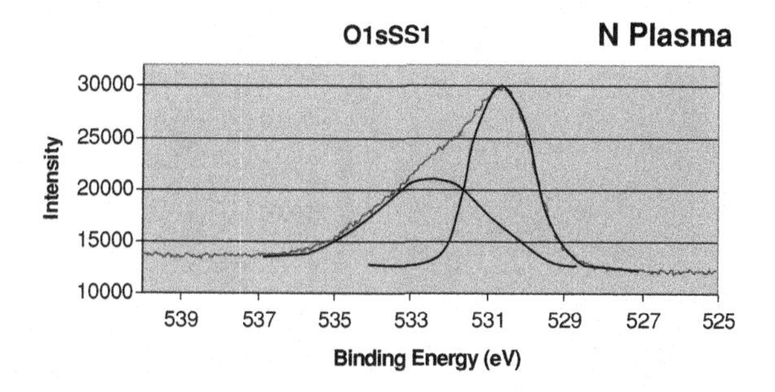

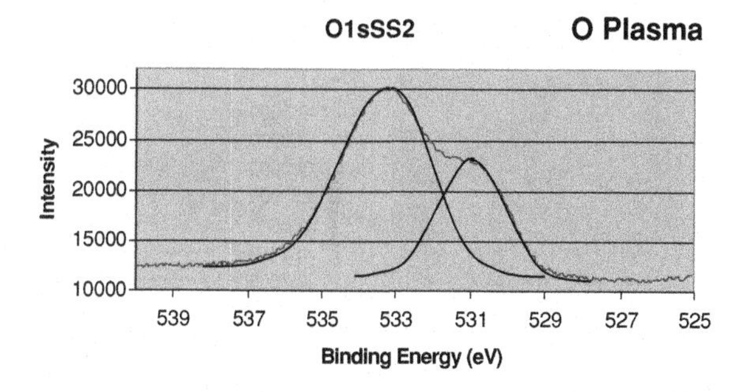

Fig. 4.21 Stainless steel XPS spectra, O 1 s region [4], Proc SPIE 2009, reprinted with permission

oxide layer, which is also evident from the C 1 s and O 1 s results. After oxygen plasma treatment, the titanium oxide peaks were significantly reduced, probably due to the pronounced oxygen layer from hydroxyl groups. Like the results for stainless steel, the nitrogen plasma was paramount in cleaning the metal surface of contaminants. The subsequent oxygen plasma chemically modified the metal surface with oxygen groups in addition to providing for a clean titanium oxide surface.

For mechanical testing of the bond strength between an embedded electrode and the (modified) EAP material, which was polymerized with the electrode(s) in place, a heavy counterweight was used while the electrode was pulled at a consistent rate until the electrode broke free and detached from the polymer. For stainless steel and for titanium, oxygen plasma treatment produced the strongest

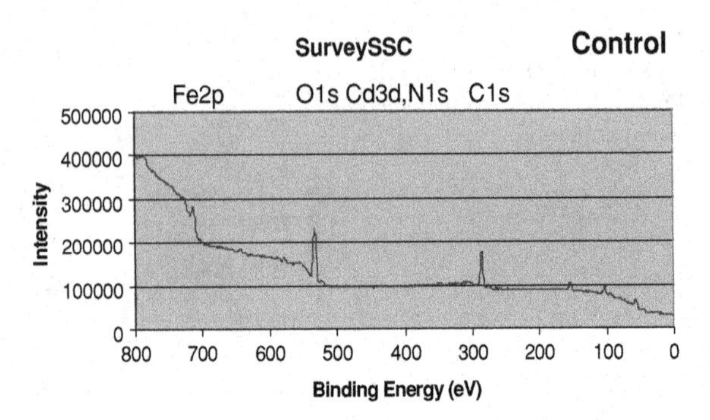

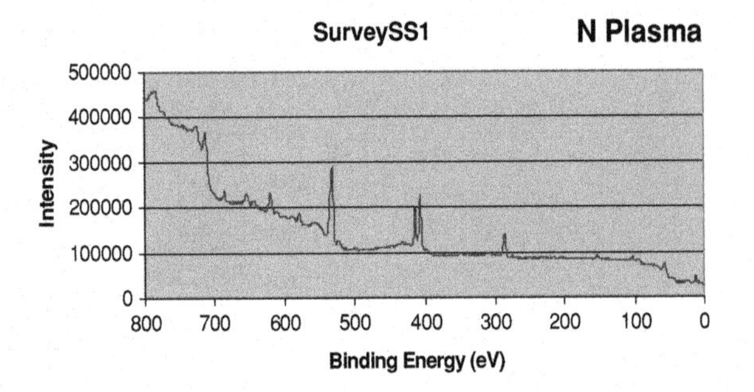

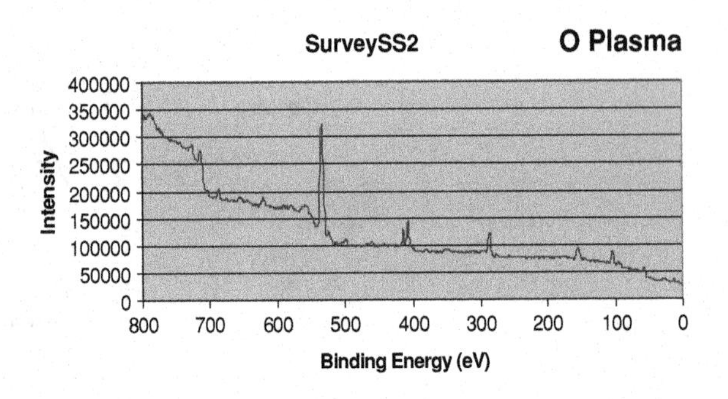

Fig. 4.22 Titanium XPS spectra [4], Proc SPIE 2009, reprinted with permission

metal–polymer interfaces (Tables 4.6 and 4.7). After the mechanical testing of the bond strength between embedded electrodes and EAP material, the metal wires were observed to determine the mode of failure using visualization and stereo microscopy. In most cases, the failure was at the interface. For a few wire samples, the mode of failure was a mixed mode between interfacial failure and failure within the polymer layer.

The polymer-metal interface between the ionic EAP and the embedded electrode(s) was significantly improved using plasma treatment. Based on the water drop surface contact angle tests and mechanical testing, oxygen plasma-treated stainless steel and titanium led to much better adhesion between the electrodes and the EAPs. For both stainless steel and titanium, XPS confirmed the presence of a

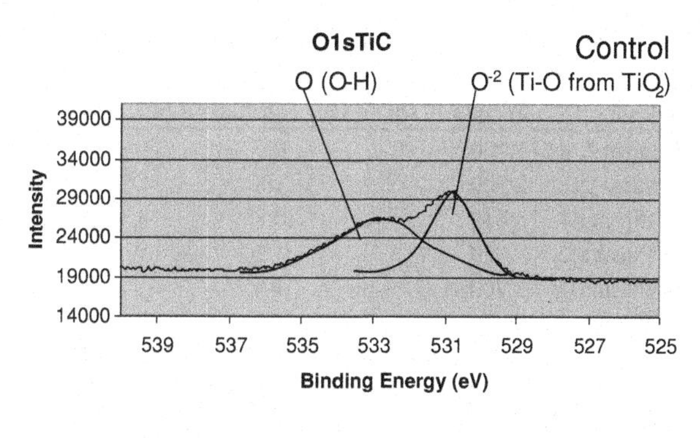

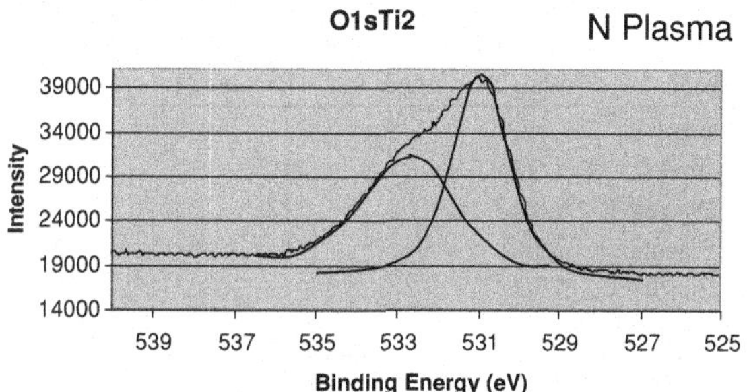

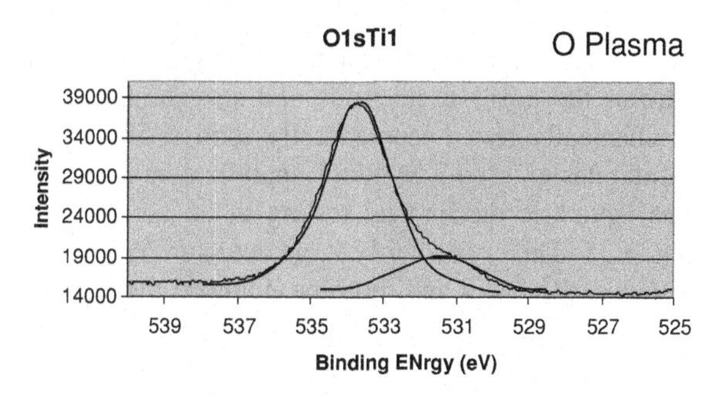

Fig. 4.23 Titanium XPS spectra, O 1 s region [4], Proc SPIE 2009, reprinted with permission

good, clean oxide layer with a significant presence of elemental oxygen in the form of hydroxyl groups after oxygen plasma treatment, which markedly increased the hydrophilicity of these metal surfaces. Other strategies to improve the polymer-metal bond are also being used in these EAP based actuators, such as configuring shaped electrodes and synthesizing multi-phasic hydrogels. In the multi-phasic materials, certain regions of the EAP comprise a much stronger, stiffer formulation that adheres tightly to selected areas of the embedded electrode, tethering it in place. The goal is for both the electroactive material and the embedded electrode to move as a unit, analogous to our muscles, nerves and tendons moving together. In this analogy, the ionic EAP based material serves as the artificial muscle, and the plasma treated electrode serves as both a tendon, connecting the EAP to a lever, and as a nerve, delivering electric stimulus to the EAP.

Table 4.6 Stress test to break of plasma treated stainless steel electrodes in EAP actuators [4], Proc SPIE 2009, reprinted with permission

Sample	Weight to break (g)	Distance (cm)	Stress ($N/m^2=kg/ms^2$)	<Stress> (N/m^2)
Control 1	1,100	0.9	122	**125**
Control 2	2,300	1.8	128	
N Plasma 1	610	0.8	76.2	**76**
N Plasma 2	920	1.2	76.8	
O Plasma 1	3,700	1	370	**349**
O Plasma 2	5,900	1.8	328	

Table 4.7 Stress test to break of plasma treated titanium electrodes in EAP actuators [4], Proc SPIE 2009, reprinted with permission

Sample	Weight to break (g)	Distance (cm)	Stress ($N/m^2=kg/ms^2$)	<Stress>(N/m^2)
Control 1[a]	183	1	18.3	**431**
Control 2	5,600	1.3	431	
N Plasma 1	6,800	1.5	453	**476**
N Plasma 2	7,500	1.5	500	
O Plasma 1	6,300	1	630	**746**
O Plasma 2	6,900	0.8	862	

[a] Control 1 not used in average data set

An ionic EAP based actuator (3.16 g), with an embedded spiral stainless steel wire as the positive electrode and an external platinum negative electrode, was mechanically tested isometrically with a 587.7 mg counterweight. The experimental design used a Mettlar® analytical balance and a pulley system (Fig. 4.24). The isometric mechanical testing of an activation-relaxation cycle is shown in Fig. 4.25. This was a fairly large actuator. We have observed that the smaller the actuator, the faster and more pronounced the contraction.

Voltage step functions (high voltage followed by low voltage) were applied to these contractile EAPs. The power supply consisted of three identical power supply units connected to provide a range of voltages: −45 V, −30 V, −15 V, 5 V, 15 V, 30 V, and 45 V. A power-modulating device was created with flip-flop integrated circuits and relays to simultaneously allow one voltage to be turned off while a different voltage is turned on. The voltage could either be maintained (control) or reduced in a step function manner. The change in size of the EAP was video-recorded and used to determine how much the EAP contracted, with the mass of the EAP determined before and after contraction (Fig. 4.26).

Figure 4.26 compares a control voltage experiment of 45 V for 40 s, for an overall contraction of 60%, to a step voltage experiment of 45 V for 10 s followed by 5 V for 30 s, for an overall contraction of 21%. The normalized results indicate that the amount of contraction is dependent on the amount of voltage applied.

Voltage reduction could be very useful for creating motor function. It is very important for muscles to be able to partially contract, exemplified by the act of holding an egg with one's fingers. If muscles could only completely contract, nobody

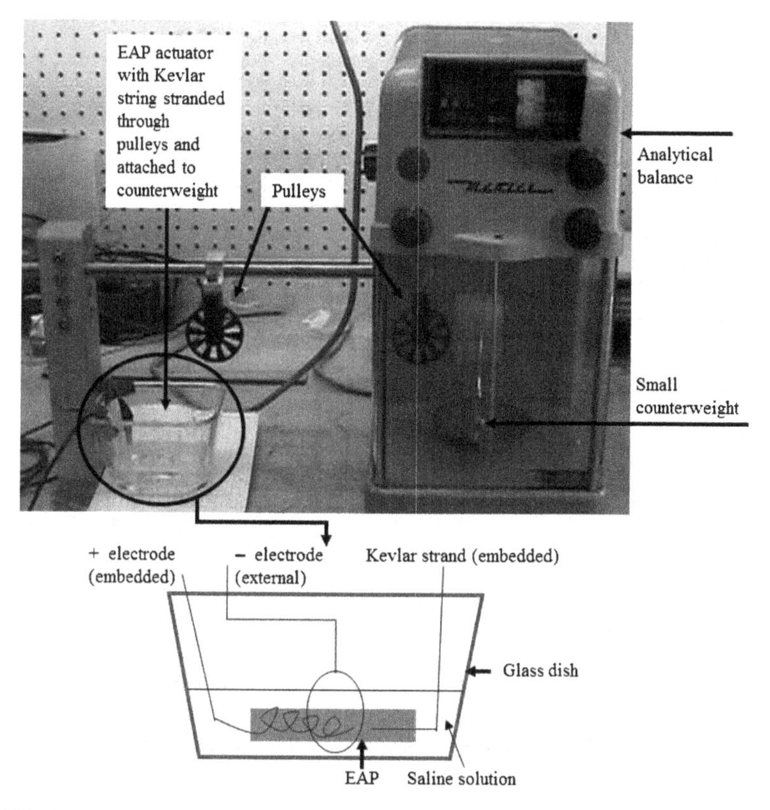

Fig. 4.24 Isometric mechanical testing set-up [5], Proc SPIE 2011, reprinted with permission

Fig. 4.25 Activation-relaxation cycle [5], Proc SPIE 2011, reprinted with permission

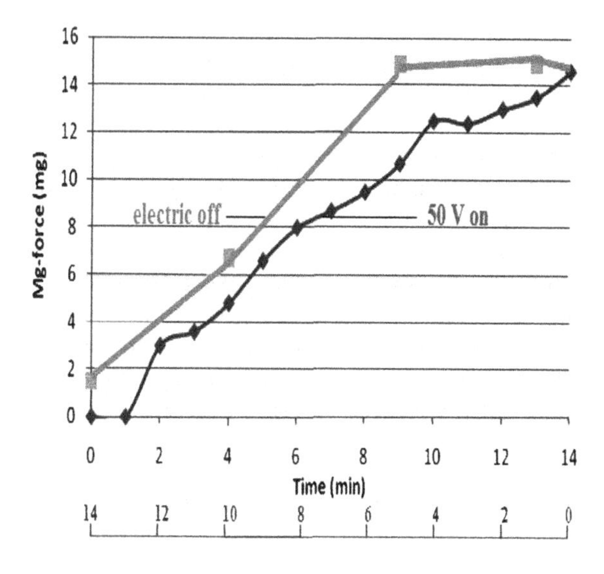

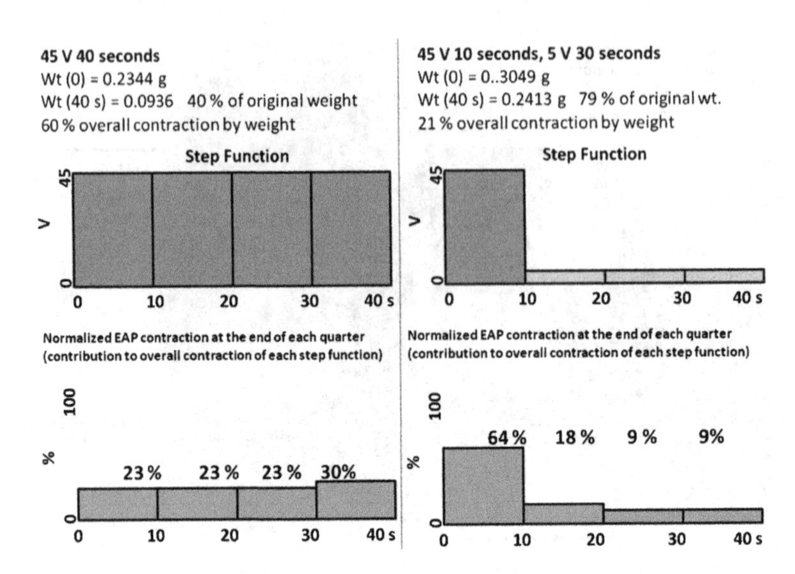

Fig. 4.26 Applied voltage step function experiments [5], Proc SPIE 2011, reprinted with permission

could hold an egg without breaking it. A combination of high and low voltages could produce gross and fine motor skills, respectively, providing both large motor control and fine motor control (fine manipulation) within the same actuator unit.

4.7 Some Thoughts Moving Forward

We have observed that the smaller the actuator, the faster and more pronounced the contraction. Moving forward, actuators using small, thin (exploring even fiber-like) contractile EAPs are being prototyped, which can then be bundled together in units to form larger, fast responding contractile actuators even more capable of life-like biomimetic movement.

Thin elastomeric coatings or coverings, which also serve as a moisture barrier, act as "skin," preventing evaporation and leakage of the electrolyte solution(s). This allows these actuators to be fully operational anywhere. Using a bilayer coating can provide for a compact actuator, where the innermost layer (the layer closer to the EAP) is a conductive layer and serves as the external electrode (Fig. 4.27).

Our muscles only contract, so in order to have pull and push movement, at the elbow for example, antagonistic pairs are used, which also provide for stability. Robust EAPs capable of fast, controlled contraction and expansion have been produced in a variety of shapes and sizes. How would we be designed if our muscles could both contract and expand (pull and push)? We would be much more streamlined for instance. These developments and concepts allow for extremely innovative designs and a new way of thinking about and configuring motion.

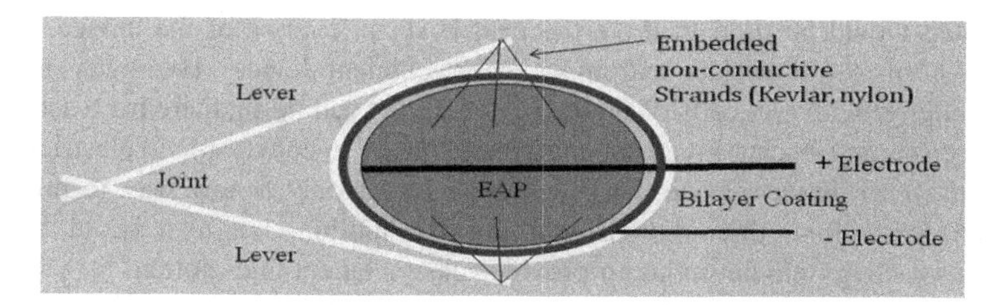

Fig. 4.27 Contractile EAP in two link actuated device

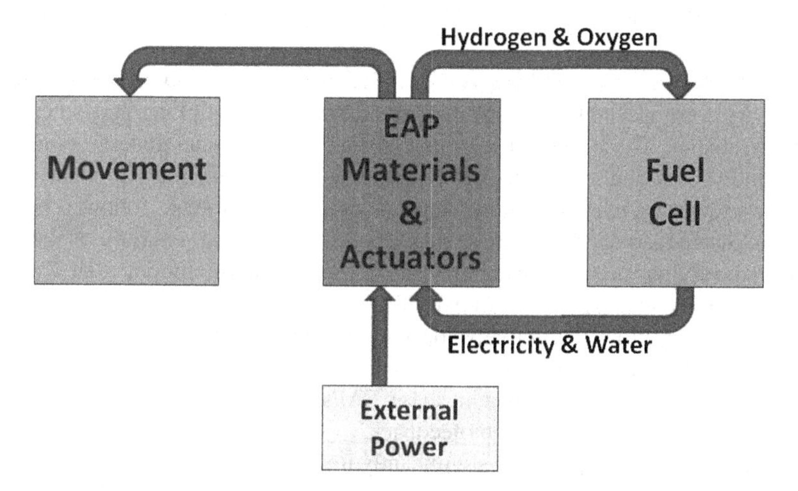

Fig. 4.28 Extremely energy efficient motion using PMA based EAPs

There are many possibilities for using electroactive materials that quickly contract and expand. Such a contracting substance could behave like the muscle tissue in the esophagus and human gut. Shaped into a series of toroid shaped rings, they could be arranged to squeeze water through a tube. That could be put to use in inventing a pump or liquid transportation system for use in remote areas. With low noise and a low heat signature, this novel pump design could be very useful for naval applications. They could also be used in robotic devices, helping to grasp loads and to climb to areas unreachable by wheeled devices. They could also activate many kinds of valves and stents, including ones used to repair damaged hearts. Medical device manufacturers could use these EAPs for controlled drug delivery and also to make catheters that first contract for easy insertion, expand once in place where the catheter is needed, and then can be contracted again for easy removal. Another application is Braille pads where a "page" could be programmed to change its configurations of Braille bumps, just as an electronic book device changes its display of pixels, making an e-reader for the visually impaired. The possibilities are as unlimited as our imagination.

Electrolytic effects are also being investigated as part of a bilayer coating. The goal is to use byproduct gases to drive a fuel cell (Fig. 4.28). The electricity generated would be used to drive the next EAP in a series of the device. Conceivably, this could produce extremely energy efficient devices. Hydrogen storage is a challenge with fuel cells. With the EAP and fuel cell tie-in, there is no need for hydrogen storage because it can be used immediately in conversion to electricity to drive the EAP based device. EAP actuation is not only a new way to provide smooth, controlled, three-dimensional life-like motion, but also could be an extremely energy efficient way to provide for movement and motion.

4.8 Conclusions

Ionic EAP based materials have been synthesized, where the strength and electroactivity were tailored by controlling the extent of crosslinking and other synthetic strategies. These EAP based materials and actuators can undergo rapid and pronounced contraction. Reversing the polarity of selected EAPs produces contraction–expansion cycles. Voltage step functions—high voltage followed by low voltage—were also applied to these contractile EAPs. Using a variety of voltages produced varying amounts of contraction. This has enormous potential for biomimetic applications. A combination of high and low voltages could produce gross and fine motor skills, respectively, providing both large motor control and fine motor control (fine manipulation) within the same actuator unit. The ability to modulate the level of movement in these EAPs by simply adjusting the voltage level could also be linked with biofeedback.

Research and development to significantly improve the polymer-metal interface has been performed, with significantly better adhesion between the electrode(s) and the EAP, particularly for titanium treated with oxygen plasma. Other strategies, such as using spiral shaped electrodes and multiphasic EAP formulations, markedly improved the connection of the embedded electrode(s) to these dynamic EAPs and improved the overall durability of the actuator. The goal is for both the electroactive smart material and the embedded electrodes to move as a unit, analogous to our muscles, nerves, and tendons moving together.

Molecular modeling and experimentation were performed to determine how and why these EAPs contract. Without electric input, the positively charged cations remain very tightly bound to the ionic EAPs. Once electricity is applied to these EAP based materials and actuators, cations and water move very rapidly from a three-fold effect due to ion migration, localized pH effects at the electrodes, and osmosis, producing a rapid pronounced contraction of the EAP. These recent developments, and fundamentally, the thorough understanding of the contraction phenomenon, are important in the field of electroactivity because of the ability of contraction and contraction–expansion to produce biomimetic life-like motion.

Acknowledgments The Department of Energy, US DOE Contract No. DE-AC02-76CHO3037 and US DOE Contract No. DE-AC02-CH0911466, are gratefully acknowledged for the partial funding of this project. Intern support was provided by the National Undergraduate Fellowship (NUF) Program and the Science, Technology, Engineering, and Mathematics (STEM) Program.

The Academies Creating Teacher Scientists (ACTS) is also gratefully acknowledged. Personal acknowledgements are gratefully expressed on pages vii–viii.

References

1. Rasmussen L, Erickson CJ, Meixler LD, Ascione G, Gentile CA, Tilson C, Abelev E (2010) Considerations for contractile electroactive polymeric materials and actuators. Polym Int 59:290–299. doi:10.1002/pi.2763
2. Rasmussen L, Schramm D, Meixler LD, Gentile CA, Ascione G, Tilson C, Pagdon K (2010) EAPAD 2010 Proc SPIE 7642: 76420Z-1–76420Z-9. doi:10.1117/12.847176
3. Rasmussen L (2009) Electroactive materials and electroactive actuators that act as artificial muscle, tendon, and skin. US Patent Application 12/319804
4. Rasmussen L, Erickson CJ, Meixler LD (2009) The development of electrically driven mechanochemical actuators that act as artificial muscle. EAPAD 2009 Proc SPIE 7287: 72871E-1–72871E-13. doi:10.1117/12.847176
5. Rasmussen L, Schramm D, Rasmussen P, Mullaly K, Meixler LD, Pearlamn D, Kirk A (2011) Considerations for contractile electroactive materials and actuators. Proc SPIE 7976: 79762B-1–79762B-13. doi:10.1117/12.880199
6. Rasmussen L (2007) Electrically driven mechanochemical artificial muscle: for smooth 3-dimensional movement in robotics and prosthetics. Proc SPIE 6524: 652423-1–652423-7. doi:10.1117/12.723273
7. Osada Y (1987) Conversion of chemical onto mechanical energy by synthetic polymers (chemomechanical systems). In: Olivé S, Henrici-Olivé (eds) Advances in Polymer Science 82. Springer, New York. doi:10.1007/PFb0024040
8. Osada Y, De Rossi DE (2010) Polymer sensors and actuators. Springer, New York
9. Osada Y, Khokhlov A (2002) Polymer gels and networks. Marcel Dekker, New York
10. Osada Y (2000) Polymer sensors and actuators. Springer, New York
11. Osada Y, Kajiwara K (2000) Gels Handbook. Academic Press, Elsevier, Amsterdam
12. Chen L, Kim B, Nishino M, Gong JP, Osada Y (2000) Environmental responses of polythiophene hydrogels. Macromolecules 2000(33):1232–1236. doi:10.1021/ma990923i
13. Narita T, Hirota N, Gong JP, Osada Y (1999) Effects of counterions and co-ions on the surfactant binding process in the charged polymer network. J Phys Chem 103:6262–6266. doi:10.1021/jp9903581
14. Uchida M, Kurosawa M, Osada Y (1995) Swelling process and order-disorder transition of hydrogel containing hydrophobic ionizable groups. Macromolecules 28:4583–4586. doi:0024.9297/95/2228-4583
15. Osada Y, Ross-Murphy SB (1993) Intelligent gels. Sci Am 268(5):82–87. doi:10.1038/scientificamerican0593-82
16. Okuzaki H, Osada Y (1992) Electro-driven chemomechanical behaviors of polymer gel based on reversible complex formation with surfactant molecules, and polymer gels: intelligent soft materials as new energy tranducers. In: Takagi T, Takahashi K, Aizawa M, Miyata S (eds) Proceeding of the first conference on intelligent materials. ICIM 92. Kanagawa
17. Osada Y, Okuzaki H, Hori H (1992) A polymer gel with electrically driven motility. Nature 355:242–243. doi:10.1038/355242a0
18. De Rossi D, Kaliwara K, Osada Y, Yamauchi A (1991) Polymer gels. Plenum Press, New York
19. Miyano M, Osada Y (1991) Electroconductive organogel 2. Appearance and nature of current oscillation under electric field. Macromolecules 24:4755–4761. doi:0024.9297/91/2224.4755
20. Osada Y (1991) Chemical valves and gel actuators. Adv Mater 3(2):107–108. doi:0935-9648/91/0202-0107

21. Gong J, Kawakami I, Osada Y (1991) Electroconductive organogel. 4. Electrodrivien chemomechanical behaviors of charge-transfer complex gel in organic solvent. Macromolecules 24:6582–6587. doi:0024.0207/91/2224.6582
22. Kishi R, Osada Y (1989) Reversible volume change of microparticles in an electric field. J Chem Soc, Faraday Trans 1 85(3):655–662. doi: 10.1039/F19898500655
23. Osada Y, Umezawa K, Yamauchi A (1988) Oscillation of electrical current in water-swollen polyelectrolyte gels. Makromol Chem 189:597–605. doi:0025-116X/88
24. Osada Y, Hasebe M (1985) Electrically activated mechanochemical devices using polyelectrolyte gels. Chem Lett 14(9):1285–1288
25. Osada Y, Sato M (1980) Conversion of chemical into mechanical energy by contractile polymers performed by polymer complexation. Polymer 21:1057–1061. doi:0032-3861/80/091057-05
26. Osada Y (1988) Electroconductive thin film. US Patent 4735852
27. Tsuchida E, Osada Y (1979) Polyion complex and method for preparing the same. US Patent 4137217
28. Tanaka T, Nishio I, Sun ST (1982) Collapse of gels in an electric field. Science 218:467–469. doi:0036-8075/82/1029-0467
29. Tanaka T, Grosberg AY (1999) Molecular dynamics of multi-chain coulomb polymers and the effect of salt ions. AIP Conf Proc 469:599–606. doi:10.1063/1.58554
30. English AE, Tanaka T, Edelman ER (1998) Polymer and solution ion shielding in polyampholytic hydrogels. Polymer 39(24):5893–5897. doi:0032-3861(98)00106-2
31. Matsuo ES, Tanaka T (1988) Kinetics of discontinuous volume-phase transistion of gels. J Chem Phys 89(3):1695–1703. doi:0021-9606/88/151695-09
32. Hirotsu S, Hirowaka Y, Tananka T (1987) Volume-phase transitions of ionized n-isopropylacrylamide gels. J Chem Phys 87(2):1392–1395. doi:0021-9606/87/141392-04
33. Tanaka T (1987) Gels. In: Nicolini C (ed) Structure and dynamics of biopolymers. Martinus Nijhoff Publishers, Boston. NATO ASI Series E
34. Hirokawa Y, Tanaka T, Sato E (1985) Phase transition of positively ionized gels. Macromolecules 18:2782–2784. doi:0024.9297/85/2218-2782
35. Tanaka T, Sato E, Hirokawa Y, Hirotsu S, Peetermans J (1985) Critical kinetics of volume phase transitions of gels. Phy Rev Lett 55(22):2455–2458. doi:10.1103/PhysRevLett.55.2455
36. Tanaka T (1985) Critical dynamics, kinetics and phase transitions of polymer gels. Polym Preprt 27(1):235
37. Tanaka T (1981) Gels. Sci Amer 244(1):124–138. doi:10.1038/scientificamerican0181-124
38. Tanaka T, Fillmore D, Sun ST, Nishio I, Swislow G, Shah A (1980) Phase transition in ionic gels. Phys Rev Lett 45(20):1636–1639. doi:10.1103/PhysRevLett.45.1636
39. Tanaka T, Nishio N, Sun ST (1992) Collapsible gel compositions. US Patent 5100933
40. Tanaka T, Hirokawa Y (1988) Reversible discontinuous volume changes of ionized isopropylacrylamide gels. US Patent 4732930
41. Shahinpoor M, Schneider HJ (2008) Intelligent materials. RSC Publishing, Cambridge
42. Shahinpoor M, Kim KJ (2005) Ionic polymer-metal composites IV. Industrial and medical applications (review paper). Smart materials and structures. Int J 14(1):197–214. doi:10.1088/0964.1726/14/1/020
43. Shahinpoor M, Kim KJ (2004) Ionic polymer-metal composites III. Modeling and simulation as biomimetic sensors, actuators, transducers and artificial muscles (review paper). Smart materials and structures. Int J 13(6):1362–1388. doi:10.1088/0964.1726/13/6/009
44. Shahinpoor M (2003) Ionic polymer-conductor composites as biomimetric sensors, robotic actuators and artificial muscles—a review. Electrochim Acta 48(14–16):2343–2353
45. Kim KJ, Shahinpoor M (2003) Ionic polymer-metal composites II. Manufacturing techniques (review paper). Smart materials and structures. Int J 12(1):65–79. doi:10.1088/0964.1726/12/1/308
46. Shahinpoor M, Kim KJ (2001) Ionic polymer-metal composites I. Fundamentals (review paper). Smart materials and structures. Int J 10(4):819–833. doi:10.1088/0964.1726/10/4/327

47. De Gennes PG, Okumura K, Shahinpoor M, Kim KJ (2000) Mechanoelectric effects in ionic gels. Europhys Lett 50:513–518. doi:10.1209/epl/i2000-00299-3
48. Shahinpoor M, Bar-CohenY, Xue T, Harrison JS, Smith J (1999) Ionic polymer-metal composites as biomimetic sensors and actuators-artificial muscles. In: Khan IM, Harrison JS (eds) Field responsive polymers. ACS Symp Series, Washington DC. doi:10.1021/bk-1999-0726.ch003
49. Segalman D, Witkowski W, Adolf D, Shahinpoor M (1992) Numerical simulation of the dynamic behavior of polymeric gels. In: Takagi T, Takahashi K, Aizawa M, Miyata S (eds) Proceedings of the first international conference on intelligent materials. ICIM 92. Kanagawa
50. Shahinpoor M, Mojarrad M (2002) Ionic polymer sensors and actuators. US Patent 6475639
51. Shahinpoor M, Mojarrad M (2000) Soft actuators and artificial muscles. US Patent 6109852
52. Shahinpoor M (1995) Spring-loaded polymeric gel actuators. US Patent 5389222
53. Bar-Cohen Y (2011) Biomimetics: nature inspired innovation. CRC Press, London
54. Bar-Cohen Y (2005) Biomimetics: biologically inspired technologies. CRC Press, London
55. Bar-Cohen Y (2004) Electroactive polymer (EAP) actuators as artificial muscles: reality, potential, and challenges, 2nd edn. PM136. SPIE Press, Bellingham
56. Bar-Cohen Y, Breazeal CJ (2003) Biologically inspired intelligent robots. PM122. SPIE Press, Bellingham
57. Bar-Cohen Y (2010) Refreshable braille displays using EAP actuators. EAPAD 2010 Proc SPIE 7642:764206-1–764206-5. doi:10.1117/12.844698
58. Bar-Cohen Y (2009) Electroactive polymer (EAP) actuators for future humanlike robots. EAPAD 2009, Proc SPIE 7287:728703-1–728703-6. doi:10.1117/12.815298
59. Bar-Cohen Y (2008) Humanlike robots as platforms for electroactive polymers. EAPAD 2008, Proc SPIE 6927:692703-1–692703-6. doi:10.1117/12.776471
60. Bar-Cohen Y (1997) Pump having pistons and valves made of electroactive actuators. US Patent 5630709
61. Andreeva AS, Philippova OE, Khokhlov AR, Islamov AK, Kuklin AI (2005) Effect of the mobility of charged units on the microphase separation in amphiphilic polyelectrolyte hydrogels. Langmuir 21:1216–1222. doi:10.1021/la0478999
62. Vasilevskaya VV, Potemkin II, Khokhlov AR (1999) Swelling and collapse of physical gels formed by associating telechelic polyelectrolytes. Langmuir 15:7918–7924. doi:10.1021/la981057q
63. Shiga T, Hirose Y, Okada A, Kurauchi T (1998) Bending of a high strength gel in an electric field. Polym Preprt 30(1):310–314
64. Zhang H, Düring L, Kovacs G, Yuan W, Niu X, Pei P (2010) Interpenetrating polymer networks based on acrylic elastomers and plasticizers with improved actuation temperature range. Polym Int 59:384–390. doi:10.1002/pi.2784
65. Yu Z, Niu X, Brochu P, Yuan W, Li H, Chen B, Pei Q (2010) Bistable electroactive polymers (BSEP): large-strain actuation of rigid polymers. EAPAD 2010, Proc SPIE 7642:76420C-1–76420C-9. doi:10.1117/12.847756
66. Brochu P, Li H, Niu X, Pei Q (2010) Factors influencing the performance of dielectric elastomer energy harvesters. EAPAD 2010, Proc SPIE 7642:76422 J-1–76422 J-12. doi:10.1117/12.847736
67. Yuan W, Brochu P, Zhang H, Jan A, Pei Q (2009) Long lifetime elastomer actuators under continuous high strain actuation. EAPAD 2009, Proc SPIE 7287:72870O-1–72870-8. doi:10.1117/12.816076
68. Lam T, Tran H, Yuan W, Yu Z, Ha S, Kaner R, Pei Q (2008) Polyaniline nanofibers as a novel electrode material for fault-tolerant dielectric elastomer actuators. EAPAD 2008 Proc SPIE 6927:69270O-1–69270O-10. doi:10.1117/12.776817
69. Yuan W, Lam T, Biggs J, Hu L, Yu Z, Ha S, Xi D, Senesky MK, Gruner G, Pei Q (2007) New electrode materials for dielectric elastomer actuators. EAPAD 2007 Proc SPIE 6524:65240 N-1–65240 N-12. doi:10.1117/12.715383

70. Ha SM, Yuan W, Pei Q, Pelrine R, Stanford S (2007) Interpenetrating networks of elastomers exhibiting 300% electrically-induced strain. Smart Mater Struct 16(2):S280–S287. doi:10.1088/0964.1726/16/2/S12
71. Ha SM, Yuan W, Pei Q, Pelrine R, Stanford S (2006) Interpenetrating polymer networks for high performance electroelastomer artificial muslces. Adv Mater 18(7):887–891. doi:10.1002/adma.200502437
72. Pei Q, Pelrine R, Stanford S, Kornbluh R, Rosenthal M (2003) Electroelastomer rolls and their application for biomimetic walking robots. Synth Met 135–136:129–131
73. Ashley S (2003) Artificial muslces. Sci Am 289(4):52–59
74. Pelrine R, Kornbluh R, Joseph J, Heydt R, Pei Q, Chiba S (2000) High-field deformation of elastomeric dielectrics for actuators. Mater Sci and Eng C 11:89–100
75. Pelrine R, Kornbluh R, Pei Q, Joseph J (2000) High-speed electrically actuated elastomers with strain greater than 100%. Science 287(5454):836–839. doi:10.1126/science.287.5454.836
76. Pei Q, Inganäs O, Lundström I (1993) Bending bilayer strips from polyaniline for artificial electrochemical muscles. Smart Mater Sructr 2(1):1–6. doi:10.1088/0964.1726/2/1/001
77. Pei Q, Inganäs O (1992) Conjugated polymers and the bending cantilever method: electrical muscles and smart devices. Adv Mater 4(4):277–278. doi:10.1002/adma.19920040406
78. Pelrine RE, Kornbluh RD, Stanford SE, Pei Q, Heydt R, Eckerle JS, Heim JR (2008) Electroactive polymer devices for moving fluid. US Patent 7362032
79. Pei Q, Pelrine RE, Kornbluh RD (2007) Electroactive polymers. US Patent 7199501
80. Pei Q, Pelrine RE, Kornbluh RD (2006) Electroactive polymers. US Patent 7049732
81. Pelrine RE, Kornbluh RD, Pei Q (2005) Electroactive polymer transducers and actuators. US Patent 6(940):211
82. Carpi F, Smela E (2009) Biomedical applications of electroactive polymer actuators. Wiley, United Kingdom
83. Balakrisnan B, Smela E (2010) Challenges in the microfabrication of dielectric elastomers. EAPAD 2010, Proc SPIE 7642:76420 K-1–76420 K-10. doi:10.1117/12.847613
84. Kujawski M, Pearse J, Smela E (2010) PDMS/graphite stretchable electrodes for dielectric elastomer actuators. EAPAD 2010, Proc SPIE 7642: 76420R-1–76420R-9. doi:10.1117/12.847249
85. Danashvar, E, Smela, E, Kipke DR (2010) Mechanical characterization of conducting polymer actuated neural probes under physiological settings. EAPAD 2010, Proc SPIE 7642:76421T-1–76421T-10. doi:10.1117/12.847694
86. Piyasena M, Shapiro B, Smela E (2009) A new EAP based on electroosmotic flow: nastic actuators. EAPAD 2009, Proc SPIE 7287:72871U-1–72871U-10. doi:10.1117/12.816138
87. Wang X, Smela E (2009) Color and volume change in PPy(DBS). J Phys Chem C 113:359–368. doi:10.1021/jp802937v
88. Wang X, Shapiro S, Smela E (2009) Development of a model for charge transport in conjugated polymers. J Phys Chem C 113:382–401. doi:10.1021/jp802941m
89. Wang X, Smela E (2009) Experimental studies of ion transport in PPy(DBS). J Phys Chem C 113:369–381. doi:10.1021/jp809092d
90. Liu Y, Gan Q, Baig S, Smela E (2007) Improving PPy adhesion by surface roughening. J Phys Chem C 111:11329–11338. doi:10.1021/jp071871z
91. Smela E, Christopherson M, Prakash SB, Urdaneta M, Dandin M, Abshire P (2007) Integrated cell-based sensors and "cell clinics" utilizing conjugated polymer actuators. EAPAD 2007, Proc SPIE 6524:65240G-1–65240G-10. doi:10.1117/12.720295
92. Smela E, Gadegaard N (2001) Volume change in polypyrrole studied by atomic force microscopy. J Phys Chem B 105:9395–9405. doi:10.1021/jp004126u
93. McKay T, O'Brien B, Calius E, Anderson I (2010) An integrated dielectric elastomer generator model. EAPAD 2010, Proc SPIE 7642:764216-1–64216-11. doi:10.1117/12.847838

94. Anderson I, Ieropoulos I, McKay T, O'Brien B, Melhuish C (2010) A hybrid microbial dielectric elastomer generator for autonomous robots. EAPAD 2010, Proc. of SPIE 7642: 7642iY-1–76421Y-11. doi:10.1117/12.847379
95. O'Brien B, Gisby T, Xie S Q, Calius E, Anderson I (2010) Biomimetic control for DEA arrays. EAPAD 2010, Proc SPIE 7642:764220-1–764220-11. doi:10.1117/12.847834
96. O'Brien B, Gisby T, Calius E, Xie S, Anderson I (2009) FEA of dielectric elastomer minimum energy structures as a tool for biomimetic design. EAPAD 2009, Proc SPIE 7287: 728706-1–728706-11. doi:10.1117/12.815818
97. Gisby T, Xie S, Calius E, Anderson I (2009) Integrated sensing and actuation of muscle-like actuators. EAPAD 2009, Proc SPIE 7287:728707-1–728707-12. doi:10.1117/12.815645
98. McKay TG, Casius E, Anderson A (2009) The dielectric COnstatn of 3 M VHB: a parameter in dispute. EAPAD 2009, Proc SPIE 7287:72870P-1–72870P-10. doi:10.1117/12.815821
99. Anderson I, Calius E, Gisby T, Hale T, McKay T, O'Brien B, Walbran S (2009) A dielectric elastomer actuator thin membrane rotary motor. EAPAD 2009, Proc SPIE 7287:72871H-1–72871H-10. doi:10.1117/12.815823
100. Walbran S, Calius E, Dunlop GR, Anderson I (2009) Optimization of electrode placement in electromyographic control of dielectric elastomers. EAPAD 2009, Proc SPIE 7287: 728724.1–728724.12. doi:10.1117/12.815833
101. O'Brien B, McKay T, Calius E, Xie S, Anderson I (2009) Finite element modelling of dielectric elastomer minimum energy structures. Appl Phys A Mater Sci Process 94(3):507–514. doi:10.1007/s00339-008-4946-8
102. O'Brien B, Calius E, Xie S, Anderson I (2008) An experimentally validated model of a dielectric elastomer bending actuator. EAPAD 2008, Proc SPIE 6927:69270T-1–69270T-11. doi:10.1117/12.776098
103. McKay T, Calius E, Anderson I (2007) Modeling a dielectric elastomer actuator based on the mckibben muscle. EAPAD 2007, Proc SPIE 6524:65241 W-1–65241 W-6. doi:10.1117/12.715842
104. Anderson IA, Kim L (2006) Force Measurement. In: Akay M (ed) Wiley encyclopedia of biomedical engineering. Wiley, New York, pp 1–4
105. Carpi F, Frediani G, Tarantino S, De Rossi D (2010) Millimetre-scale bubble-like dielectric elastomer actuators. Polym Int 59:407–414. doi:10.1002/pi.274
106. Gallone G, Galantini F, Carpi F (2010) Perspectives for new dielectric elastomers with improved electromechanical actuation performance: composites versus blends. Polym Int 59:400–406. doi:10.1002/pi.2765
107. Carpi F, Frediani G, De Rossi D (2010) Hydrostatically coupled dielectric elastomer actuators for tactile displays and cutaneous stimulators. EAPAD 2010, Proc SPIE 7642: 76420E-1–76420E-6. doi:10.1117/12.847562
108. Carpi F, Frediani G, De Rossi D (2009) Dielectric elastomeric actuators with hydrostatic coupling. EAPAD 2009, Proc SPIE 7287:72870D-1–72870D-6. doi:10.1117/12.815459
109. Carpi F, De Rossi D, Kornbluh R, Pelrine R, Sommer-Larsen P (2008) Dielectric elastomers as electromechanical transducers: fundamentals, materials, devices models and applications of an emerging electroactive polymer technology. Elsevier Ltd, Oxford
110. Carpi F, Mannini A, De Rossi D (2008) Elastomeric contractile actuators for hand rehabilitation splints. EAPAD 2008, Proc SPIE 6927:692705-1–692705-10. doi:10.1117/12.774644
111. Carpi F, Gallone G, Galantini F, De Rossi D (2008) Enhancement of the electromechanical transduction properties of a silicone elastomer by blending with a conjugated polymer. EAPAD 2008, Proc SPIE 6927:692707-1–692707-11. doi:10.1117/12.776641
112. Carpi F, De Rossi D (2007) Contractile folded dielectric elastomer actuators. SPIE electroactive polymers and devises (EAPAD) 2007, Proc SPIE 6524:65240D-1–65240D-13. doi:10.1117/12.715594
113. Carpi F, Salaris C, De Rossi D (2007) Folded dielectric elastomer actuators. Smart Mater Struct 16(2):S300–S305. doi:10.1088/0964.1726/16/2/S15

114. Carpi F, Migliore A, Serra G, De Rossi D (2005) Helical dielectric elastomer actuators. Smart Mater Struct 14(6):1210–1216. doi:10.1088/0964.1726/14/6/014
115. Carpi F, De Rossi D (2004) Dielectric elastomer cylindrical actuators: electromechanical modeling and experimental evaluation. Mater Sci Eng C 24(4):555–562. doi:10.1016/j.msec.2004.02.005
116. Carpi F, Chiarelli P, Mazzoldi A, De Rossi D (2003) Electromechanical characterization of dielectric elastomer planar actuators: comparative evaluation of different electrode materials and different counterloads. Sens Actuators A Phys 107(1):85–95. doi:10.1016/S0924.4247(03)00257-7
117. Mirfakhrai T, SHoa T, Fekri N, Madden J D (2009) Electrically-activated catheter using polypyrrole actuators: cycling effects. EAPAD 2009, Proc SPIE 7287:72871I-1–72871I-9. doi:10.1117/12.816056
118. John S, Aloci G, Spinks G, Madden J D, Wallace G (2008) Sensor response of polypyrrole trilayer benders as a function of geometry. EAPAD 2008, Proc SPIE 6927:692721-1–692721-9. doi:10.1117/12.776171
119. Shoa T, Cole M, Munce NR, Yang V, Madden JD (2007) Polypyrrole operating voltage limits in aqueous sodium hexafluorophospate. EAPAD 2007, Proc SPIE 6524:652421-1–652421-8. doi:10.1117/12.715072
120. Lacour LP, Prahlad H, Pelrine R, Wagner S (2004) Mechatronic system of dielectric elastomer actuators addressed by thin film photoconductors on plastic. Sens Actuators A Phys 111(2–3):288–292. doi:10.1016/j.sna.2003.12.009
121. Pelrine R, Rosenthal M, Meijer K (2002) Dielectric elastomer artificial muscle actuators: toward biomimetric motion. EAPAD 2002, Proc SPIE 4695:126–137
122. Pelrine RE, Kornbluh RD, Eckerle JS (2005) Energy efficient electroactive polymers and electroactive polymer devices. US Patent 6911764
123. Kornbluh RD, Pelrine RE (2005) Variable stiffness electroactive polymer systems. US Patent 6882086
124. Pelrine RE, Kornbluh RD (2004) Non-contact electroactive polymer electrodes. US Patent 6707236
125. Pelrine RE, Kornbluh RD, Oh S, Joseph JP (2003) Electroactive polymer fabrication. US Patent 6543110
126. Rasmussen L (1998) Process for producing an electrically driven mechanochemical actuator. US Patent 5736590
127. Masiak M, Hyk W, Stojek Z, Ciszkowska M (2007) Structural changes of polyacids initiated by their neutralization with various alkali metal hydroxides. Diffusion studies in poly(acrylic acid)s. J Phys Chem B 111:11194–11200. doi:10.1021/jp0711904
128. Lide DR (2008) CRC Handbook of chemistry and physics, 89th edn. CRC Press, New York
129. Wu Y, Joseph S, Aluru NR (2009) Effect of cross-linking on the diffusion of water, ions, and small molecules in hydrogels. J Phys Chem B 113:3512–3520. doi:10.1021/jp808145x
130. Mudry B, Guy RH, Delgado-Charro MB (2006) Transport numbers in transdermal iontophoresis. Biophy J 90:2822–2830. doi:10.1529/biophysj.105.074609

Chapter 5
Touch Sensitive Dielectric Elastomer Artificial Muscles

Todd Gisby, Ben O'Brien and Iain A. Anderson

Abstract Comb jellies are tiny sea animals that do not have brains, yet they can control the synchronous beat of hundreds of swimming paddles to navigate the water column in search of food. Waves of actuation travel down rows of paddles that run the length of the animal's body to generate thrust. This is achieved using distributed local feedback and a simple control rule: each paddle only actuates when it is touched, and when it actuates it sweeps forward to touch the next paddle in line. No central brain is required to tell each paddle when to fire. We have created a scalable array of Dielectric Elastomer Actuators (DEA) that mimics the swimming paddles of the comb jelly and have implemented this array in a simple conveyor mechanism. Each DEA is made touch sensitive by sensing changes in its capacitance, eliminating the need for bulky external sensors. The array is inherently self-regulating and each DEA only actuates when it is touched, ensuring the conveyor automatically adjusts to the properties of the object being conveyed. This is a simple solution to a simple application, but it brings us one step closer to scalable, artificial muscle actuator arrays that might perform such useful tasks as assembly line conveyance and water propulsion. It also paves the way for more advanced systems that take into account DEA properties other than capacitance such as electrode resistance and leakage current.

Keywords Dielectric elastomer · DEA · DEMES · Actuator · Comb jelly ctenophore · Conveyor · Self-sensing

T. Gisby · B. O'Brien · I. A. Anderson
Biomimetics Laboratory, Auckland Bioengineering Institute, Auckland, New Zealand

I. A. Anderson (✉)
Department of Engineering Science, University of Auckland, Auckland, New Zealand
e-mail: i.anderson@auckland.ac.nz

L. Rasmussen (ed.), *Electroactivity in Polymeric Materials*,
DOI: 10.1007/978-1-4614-0878-9_5,
© Springer Science+Business Media New York 2012

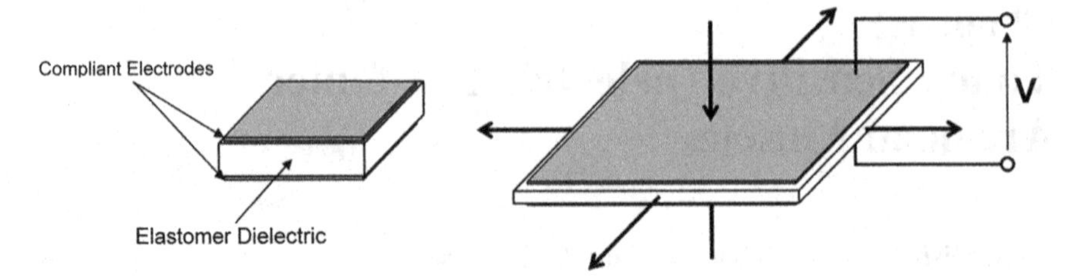

Fig. 5.1 Actuation process of DEA [2] Sensors and Actuators 1998, reprinted with permission

We rely on our sense of touch to manipulate objects and through proprioception, have the ability to judge the position of our limbs and fingers. These capabilities are made possible due to strain sensing nerves within muscle and other body tissues. For example, humans can adjust a "blind" nut on the back panel of an engine, or touch their nose with their eyes closed, capabilities that would be impossible without body tissues which have fully integrated strain sensing. Imparting such a sensing capability to electroactive polymer actuators, such as the Dielectric Elastomer Actuator (DEA), would bring the scientific community closer to producing a true engineering analog to a natural actuator system like the hand and arm.

Robotic devices can be given the ability to sense strain through the attachment of thin, surface mounted strain gauges. These gauges use the strain-dependent resistance of a thin wire or patterned metallic coating [1] and are sensitive to strains as small as thousandths of a percent. However, standard strain gauges are unsuited for attachment to the soft, low modulus polymer surfaces of DEA artificial muscles which are capable of producing strains in excess of 300% [2–4]. With few, if any, options available, it is necessary to look to the artificial muscles themselves for strain sensing, which they can provide through self-sensing. Success would impart the proprioceptive feedback needed for position control; this can be used to impart touch sensitivity. But first, a measurable parameter must be found that can also be directly related to the state of strain of the actuator. This parameter may be determined by consideration of how DEAs operate.

A DEA is a stretchable capacitor consisting of a soft polymer membrane dielectric with compliant electrodes on both faces. The charge accumulated on the electrodes, after a voltage is applied, gives rise to electrostatic forces that generate deformation in the DEA [2–4]. Charges of opposite polarity act to draw the positive and negative electrodes together while charges of the same polarity expand the area of the electrode (Fig. 5.1). When the charge is removed, the elastic energy stored in the dielectric returns it to its original shape.

Where the thickness of the dielectric membrane is approximately uniform and is much smaller than either of the planar dimensions, the DEA can be modelled as a parallel plate capacitor by Eq. 5.1, where C is capacitance, ε_r is the relative permittivity of the membrane material, ε_0 is the permittivity of free space, A is the area of the electrodes, and d is the thickness of the membrane. The permittivity of free space is constant; therefore, the capacitance of a DEA is related to the relative permittivity of the membrane, A, and d.

$$C = \varepsilon_r \varepsilon_o \frac{A}{d} \tag{5.1}$$

In tests where good contact has been ensured between the electrodes and the dielectric membrane, it has been shown that the relative permittivity of several common DEA membrane materials exhibit little to no stretch dependence over very large stretches [5–7]. Furthermore, soft elastomers can typically be regarded as being volumetrically incompressible because their bulk modulus is several orders of magnitude greater than their shear modulus [8]. Thus, when a DEA changes shape, either through electrostatic stresses or external loads, its volume is conserved: an increase in area A must be accompanied by a complimentary decrease in thickness d. A 5% increase in area A will produce a reciprocal reduction in d, leading to a 10.25% increase in capacitance. Detecting capacitance change therefore provides a potentially sensitive means of detecting electrically or mechanically induced strains in the DEA.

To self-sense the capacitance of a DEA, it is necessary to understand how it can be interrogated without affecting its function as an actuator. Consider the following simple electrical analysis of a DEA [9]. As part of a circuit, it is straightforward to directly measure the series current (i_{series}) through, and the voltage across (V_{DEA}) a DEA. Estimating the capacitance of a DEA (C_{DEA}) while it is being actuated therefore requires finding a way of interpreting the effects of capacitance on these measurable electrical signals. In addition to capacitance, however, a DEA has two other key electrical parameters that can vary depending on the operating conditions: the equivalent parallel resistance of the dielectric membrane (R_{EPR}), and the equivalent series resistance of the electrodes (R_{ESR}) (Fig. 5.2). Each of these parameters will also have a measurable effect on i_{series} and V_{DEA}. It is important therefore to consider what the i_{series} and V_{DEA} signals represent in order to extract information regarding capacitance.

The series current through a DEA is equal to the sum of the current through C_{DEA} (i_C) and leakage current through R_{EPR} (i_{EPR}). For a variable capacitor, i_C, is equivalent to the first derivative with respect to time of the charge, thus in the case of the DEA, it is the first derivative with respect to time of $C_{DEA}V_C$, taking care to note both C_{DEA} and V_C potentially have non-zero derivatives. The series current is therefore the sum of three components: the current due to a changing applied voltage, $C_{DEA}(dV_C/dt)$; the current due to a changing capacitance, $V_C(dC_{DEA}/dt)$; and the current due to leakage through the membrane, i_{EPR} (Eq. 5.2). The voltage difference between the terminals of the DEA (V_{DEA}) is equal to the sum of V_C and the product of i_{series} and R_{ESR} (Eq. 5.3).

$$i_{series} = C_{DEA}\frac{dV_C}{dt} + V_C\frac{dC_{DEA}}{dt} + i_{EPR} \tag{5.2}$$

$$V_{DEA} = i_{series}R_{ESR} + V_C \tag{5.3}$$

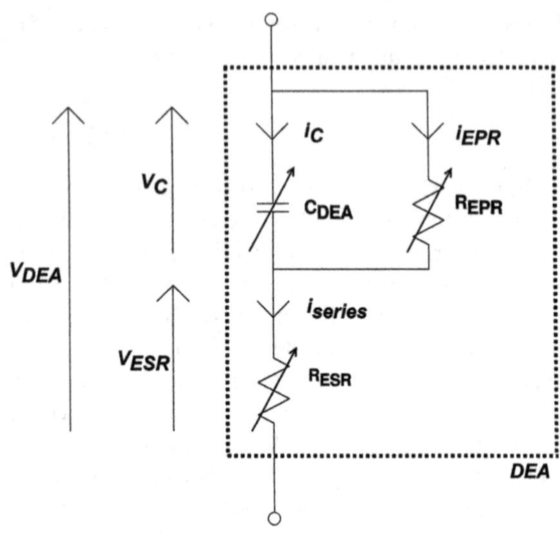

Fig. 5.2 A lumped circuit model of a DEA

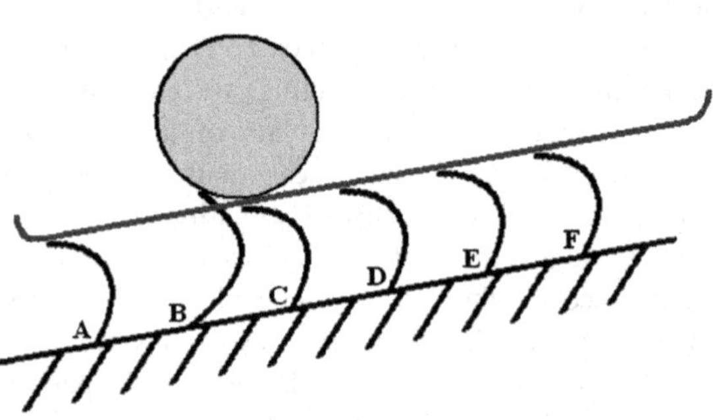

Fig. 5.3 A multi-actuator conveyor pushing a cylinder [11]. Proceedings of SPIE 2009, reprinted with permission

Equations 2 and 3 are the general equations that form the foundation upon which self-sensing is based. However, there are six unknowns in these two equations: C_{DEA}, dC_{DEA}/dt, V_C, dV_C/dt, i_{EPR}, and R_{ESR}. Additional steps are required to reduce this system to one that is solvable. There are many approaches to take in order to do so, and will to a large extent depend on the specific application. In certain instances it is possible to greatly simplify Eqs. 5.2 and 5.3 to obtain useful capacitance feedback.

As an example, consider the design of a simple conveyor mechanism, depicted in Fig. 5.3, built from multiple DEAs that will push a small cylindrical object forward. One solution is to use a central controller to coordinate the behavior of each actuator. Without feedback, the timing of actuation of each DEA would need to follow a predetermined pattern. However, as the object accelerates, or if objects with different masses or rotational inertias are transported, it would be difficult to ensure synchronized actuation in such an inflexible system. To improve flexibility, feedback from each actuator is required. Typically this would involve conventional external sensors, e.g., laser displacement sensors, LVDTs, optical encoders, or motion sensors, and the use of a centralized controller to coordinate the overall

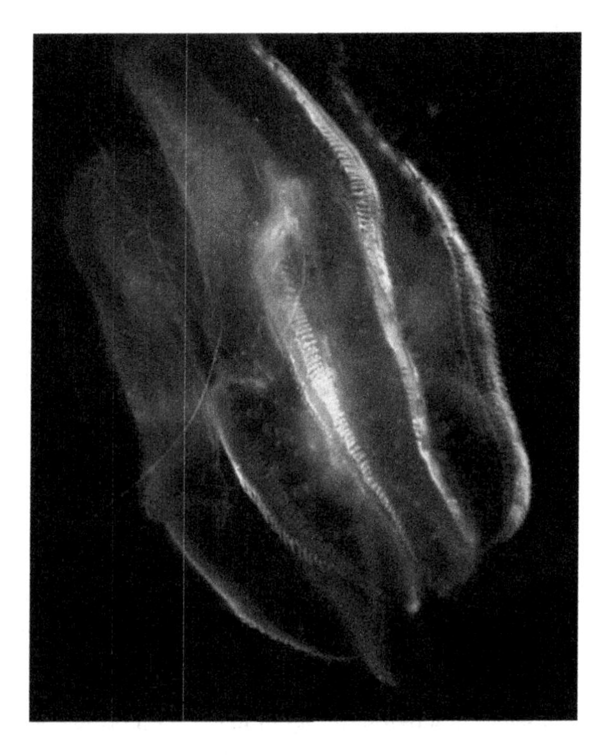

Fig. 5.4 A photo of a lobate ctenophore (*Leucothea* sp.) taken at the Poor Knights Islands, NZ (Iain A. Anderson)

behavior of the system. However, such a system would grow substantially in complexity, size, and cost as actuators are added.

Clearly a conventional approach where actuation and sensing functions are performed by distinct components and are coordinated using a centralized controller is not optimal when reduced component count and scalability are important. However, nature can be the inspiration to produce alternative solutions: ctenophora, or comb jellies, have been in existence for almost half a billion years (Fig. 5.4) and live in the upper ocean where they consume zooplankton that includes small crustacea, jellyfish, and each other [10]. These neutrally buoyant and delicate creatures do not have a brain, yet they can control the synchronous beat of hundreds of swimming paddles, each composed of numerous extra-long cilia, that are arranged in eight longitudinal rows. Thus, for directions on how to control a multi-actuator system with low computational overheads we can look to mimicking the mechanism and control strategy used by the comb jelly.

When activated the cilia in a comb paddle beat in unison. The paddle swings around fast and stiff during the power stroke, while in the return stroke it becomes flaccid, like a piece of overcooked spaghetti, and rolls backward (Fig. 5.5). The active paddle, in turn, mechanically triggers its neighbour, and like a row of falling dominos, the paddles in a row are actuated one behind the other. This generates forward thrust for the animal. By sending an increased number of waves of actuation along one side, the comb jelly can produce more thrust on one side and

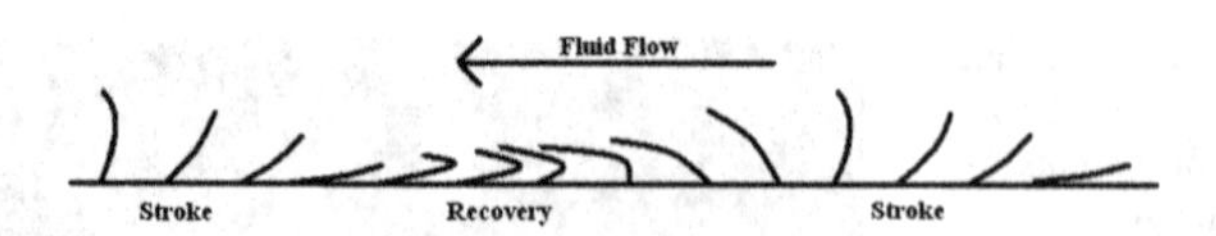

Fig. 5.5 When a ctenophore plate beats, its movement is qualitatively similar to a swimmer's arm doing the breaststroke. A power stroke pushes fluid backwards and is followed by a slow recovery stroke [11]. Proceedings of SPIE 2009, reprinted with permission

steer itself. This enables the animal to maintain a vertical orientation and to swim up or down in the water column to intercept the plankton. The triggering of each row is controlled by a balance organ at the back end of the animal.

It is clear that there is enormous complexity within each ctenophore plate; it would be a herculean task to reproduce the molecular motor of each cilium and to assemble literally millions of these mechanisms into a ctenophore-like swimming device. It is possible, however, to distill and mimic key features of the actuator and control system:

1. Each paddle acts directly without transduction of force or motion, eliminating the need for complex transmission systems that, for example, convert rotary motion to linear motion, or increase/reduce speed through an intermediary device such as a gearbox or crank assembly.
2. Each paddle actuator responds only when influenced by its neighbour. So the system is self-controlling in that a push occurs at the appropriate time, greatly simplifying the task of coordination.
3. There is no central brain. Instead there is a simple control strategy at work that results in a coordinated and useful response: forward thrust for the animal.
4. Multiple paddles and rows provide substantial system redundancy.

Returning to the conveyor system, consider mimicking the comb jelly by making each actuator autonomous, and giving each actuator touch sensitivity, thus eliminating the need for central control. First, however, a suitable dielectric elastomer device that can be made touch sensitive is required. A good candidate for this is the Dielectric Elastomer Minimum Energy Structure (DEMES) [11, 12]. A DEMES can be fabricated by stretching a dielectric membrane, in tension, across a window cut within a thin flexible plastic frame. When released, tension in the membrane causes the frame to buckle. By holding down one edge of the frame, the buckling will make the DEMES stand up (Fig. 5.6). By applying a flexible electrode onto the faces of the membrane within the window of the plastic frame, a DEA can be produced. Actuation of the DEA membrane will relieve tension within it, causing the DEMES to revert to its flat state. Thus a bending actuator is created, the tip of which can be used to push objects around.

Starting from its curled up rest state (Fig. 5.6, far right image), pushing the tip of the DEMES forward or backward will change the membrane capacitance. Pushing its tip backward and thereby increasing the angle of bending will further relieve tension in the membrane. This will reduce the membrane area and increase its thickness, resulting in a reduced capacitance. Increasing the tension in the

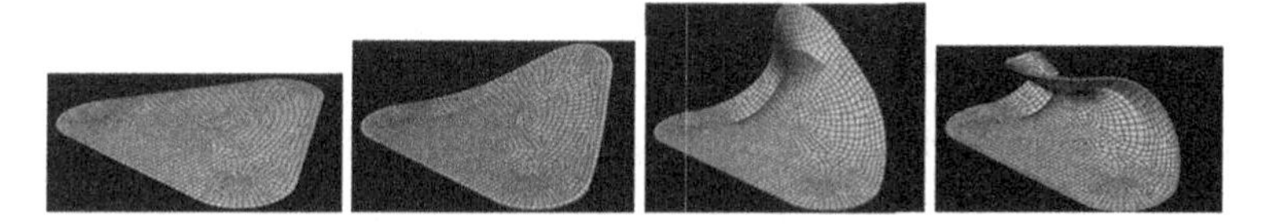

Fig. 5.6 A computer model of a DEMES actuator at several activation states. Full activation can cause the DEMES to lie flat (*far left*). No voltage across the membrane window will cause it to stand up (*far right*) [11]. Proceedings of SPIE 2009, reprinted with permission

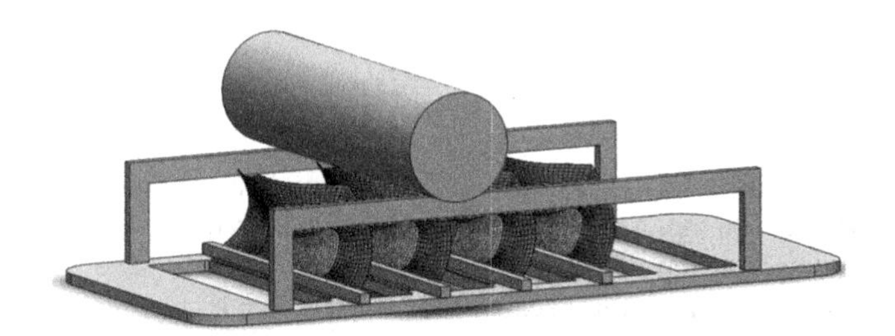

Fig. 5.7 A CAD representation of a DEMES conveyor. The object that will be supported is on the two rails [11]. Proceedings of SPIE 2009, reprinted with permission

membrane by pushing the tip forward will increase the membrane area and reduce the thickness, causing a rise in capacitance. Therefore, if capacitance change in the DEMES membrane can be sensed, it can be made touch sensitive [11], enabling the ability to detect contact between the DEMES tip and another actuator or object. Therefore, the swimming action of the comb jelly can be reproduced; ensuring each DEMES will only turn on when the previous actuator passes it the object. The conveyor will automatically adapt to the properties and speed of the object, eliminating the need for a centralized controller to coordinate each DEMES.

In Fig. 5.7 an example of a conveyor mechanism consisting of multiple artificial muscle actuators pushing a small cylindrical object up some inclined rails is shown. To realise our ctenophore-like DEMES conveyor system, however, a method is needed for measuring capacitance.

Recalling Eqs. 5.2 and 5.3, there are a number of potential phenomena that will have an effect on the electrical signals coming from the DEA. However, the nature of the conveyor application means we can greatly simplify these equations and still achieve the desired response. In particular, each DEMES is stationary and is at a relatively low voltage when in its rest state, thus we can assume both the current induced due to the rate of change of capacitance and leakage current are negligible. Furthermore, only the capacitance of the DEA needs to be polled to detect when it changes as the preceding actuator perturbs it, thus the effects of the electrode resistance can also be ignored. We can therefore approximate V_C with V_{DEA}, simplify Eq. 5.2 to Eq. 5.4, and solve for C_{DEA},

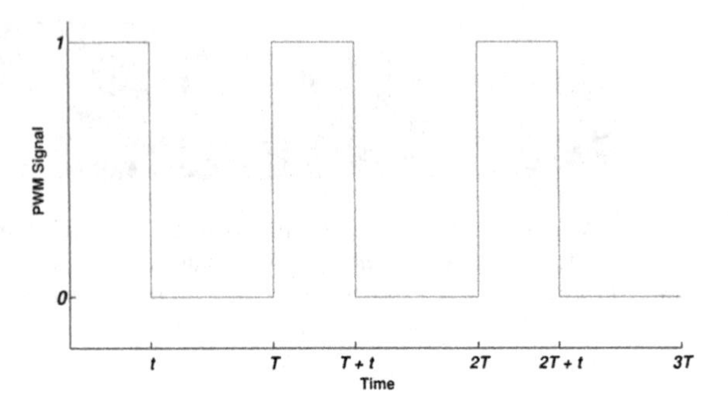

Fig. 5.8 Schematic diagram of a PWM waveform. The period of the PWM signal is equal to *T*, and the duty cycle of the PWM signal is equal to *t/T*

$$i_{series} = C_{DEA} \frac{dV_{DEA}}{dt} \tag{5.4}$$

Clearly to evaluate C_{DEA} using Eq. 5.4, a time varying voltage signal is required. Several examples of superimposing a high frequency, low amplitude sinusoidal voltage on top of the actuation voltage of a DEA have been used to achieve this [13–15]. However, because the DEMES in our conveyor example are small and do not require high power, and because faithful reproduction of a high frequency signal is difficult using typical high voltage/low current power supplies that have rectified outputs, Pulse Width Modulation (PWM) was used instead [9]. A PWM signal is a square wave with a fixed frequency and an adjustable duty cycle, where the duty cycle is the ratio of the time the signal is high to the period of the signal (see Fig. 5.8). By making the period, *T*, of the PWM signal sufficiently small relative to the electrical and mechanical time constants of the DEA, controlling the duty cycle of the signal (*t/T*) controls the average voltage across the DEA, which governs the degree of actuation. At the same time, the rapid switching of the PWM signal introduces small scale oscillations to this voltage that enable C_{DEA} to be estimated. PWM has a number of inherent benefits: it is a digital technique that is readily compatible with digital computing; it does not need a separate circuit to generate the small scale oscillations; and it enables the power supply to be set to a fixed voltage and used to power multiple independent DEAs.

Using PWM, it is straightforward to measure the average value of i_{series} and dV_{DEA}/dt using a simple voltage divider (Fig. 5.9) during the "off" portion of a PWM cycle, i.e., when no current is flowing between the DEA and the power supply. C_{DEA} can therefore be estimated using Eq. 5.4. This self-sensing can now be applied to the DEMES conveyor application. The desired behaviour of the conveyor is simple: each DEMES should turn on when it detects the touch of the previous actuator, and turn off a short time later. The conveyor therefore does not require a centralized controller to coordinate the actuators; rather, each DEMES must simply wait for its trigger and actuate if it receives it. The traveling wave of actuation is therefore an emergent behavior. This overall structure is well suited to the characteristics of a state machine. Importantly, by limiting the number of discrete states each DEMES can attain and eliminating any centralized controller, the system is inherently easy to scale to larger numbers of actuators.

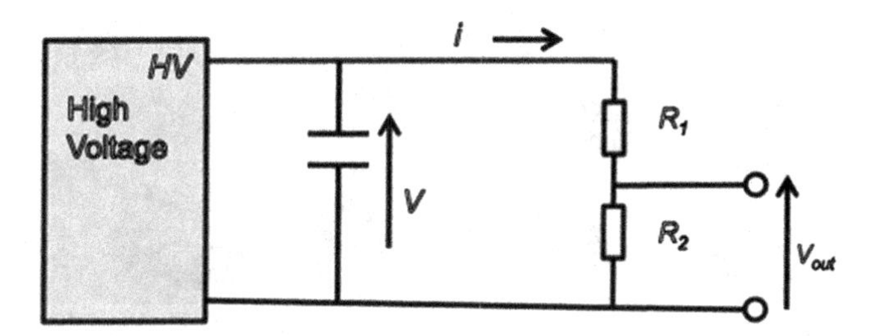

Fig. 5.9 A simple DE self-sensing circuit [11]. Proceedings of SPIE 2009, reprinted with permission

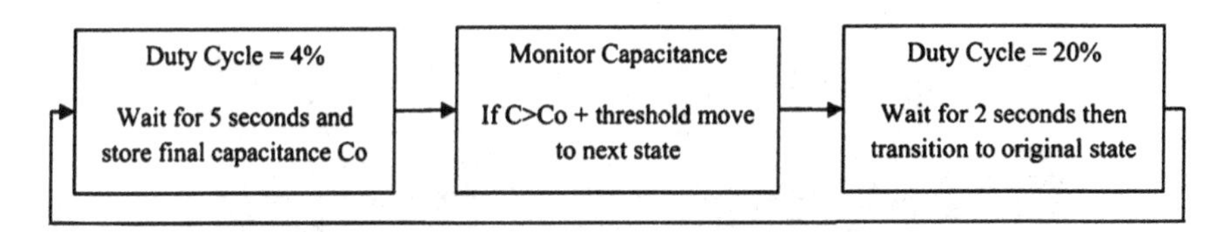

Fig. 5.10 Schematic of state machine

Conveyance of an object using our example system requires only three states for this state machine (Fig. 5.10). In the first state, a PWM signal with a low duty cycle is applied. This results in negligible deformation of the DEMES, but provides the necessary voltage oscillations to enable self-sensing. After 5 s, the estimated rest capacitance of the DEMES is recorded. The capacitance is then polled; if the capacitance increases above the rest capacitance by an appreciable amount, the system registers the contact of the preceding DEMES. Once contact has occurred, a step increase in the duty cycle is applied for a short period ($\sim$2 s), causing the DEMES to actuate. At the end of this short period, the duty cycle is reverted back to its initial value, and the DEMES re-enters the first state. Thus, when the first DEMES is triggered manually, a wave of actuation travels along the DEMES array. As a result, using self-sensing and a simple state machine, the behaviour of the comb jelly's paddles can be mimicked.

Fig. 5.11 shows an example of the DEMES conveyor in operation, pushing a small polyethylene rod along some rails.

This chapter describes the basics of capacitive self-sensing and applies it to the design of a conveyor device, actuated by DEs. Self-sensing, coupled with a simple control rule based on touch sensitivity inspired by the comb jelly, can be used to create a scalable, biomimetic actuator array. However, this is a simple solution for a simple application. For more advanced applications, particularly where continuous, proportional control of the artificial muscle element is desirable, or positional accuracy is important, it is necessary to account for the current associated with the rate of change of capacitance (the second right hand side term in Eq. 5.2). At high electric fields and large strains, it is also necessary to compensate for leakage current (the third term in Eq. 5.2) and the resistance of the electrodes. This has been comprehensively explored by Gisby [9].

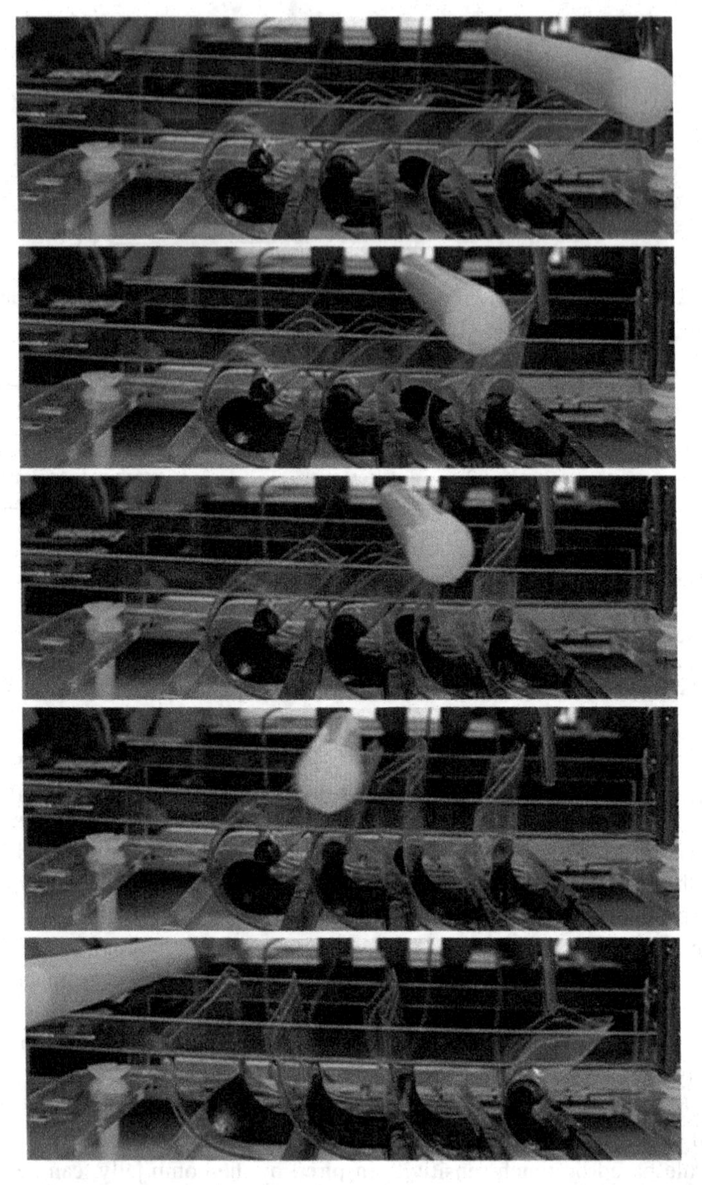

Fig. 5.11 Still images showing the Teflon roller being rolled up the rail [11]. Proceedings of SPIE 2009, reprinted with permission

We envisage a future where artificial muscle actuator arrays can be used to perform useful tasks such as product conveyance or perhaps water propulsion. They will be simple in construction without the burden of central controllers or added sensors and this will be made possible through touch sensitivity.

References

1. Anderson IA, Kim L (2006) Force measurement. In: Akay M (ed) Wiley encyclopedia of biomedical engineering. Wiley, New York, pp 1–4
2. Pelrine RE, Kornbluh RD, Joseph JP (1998) Electrostriction of polymer dielectrics with compliant electrodes as a means of actuation. Sens Actuators A: Phys 64(1):77–85. doi:10.1016/S0924-4247(97)01657-9
3. Pelrine R, Kornbluh R, Pei Q, Joseph J (2000) High-speed electrically actuated elastomers with strain greater than 100%. Science 287(5454):836–839. doi:10.1126/science.287.5454.836
4. Madden JDW et al (2004) Artificial muscle technology: physical principles and naval prospects. IEEE J Ocean Eng 29(3):706–728
5. Kofod G, Sommer-Larsen P, Kornbluh R, Pelrine R (2003) Actuation response of polyacrylate dielectric elastomers. J Intell Mater Syst Struct 14(12):787–793. doi:10.1177/104538903039260
6. Kofod G, Sommer-Larsen P (2005) Silicone dielectric elastomer actuators: finite-elasticity model of actuation. Sens Actuators A: Phys 122(2):273–283. doi:10.1016/j.sna.2005.05.001
7. McKay TG, Calius E, Anderson IA (2009) The dielectric constant of 3M VHB: a parameter in dispute. EAPAD 2009 Proc SPIE 7287:72870P. doi:10.1117/12.815821
8. Boyce MC, Arruda EM (2000) Constitutive models of rubber elasticity: a review. Rubber Chem Technol 73(3):504–523
9. Gisby TA (2011) Smart artificial muscles bioengineering. PhD Thesis. University of Auckland, NZ
10. Buchsbaum R, Buchsbaum M, Pearse J, Pearse V (1987) Animals without backbones, 3rd edn. The University of Chicago Press, Chicago
11. O'Brien B, Gisby T, Calius E, Xie S, Anderson I (2009) FEA of dielectric elastomer minimum energy structures as a tool for biomimetic design. Proc. SPIE 7287:728706-1-728706-11. doi:10.1117/12.815818
12. Kofod G, Wirges W, Paajanen M, Bauer S (2007) Energy minimization for self-organized structure formation and actuation. Appl Phy Lett 90(8):081916-1–081916-3. doi:10.1063/1.2695785
13. Toth LA and Goldenberg AA (2002) Control system design for a dielectric elastomer actuator: the sensory subsystem. EAPAD 2002 Proc SPIE 4695:323. doi:10.1117/12.475179
14. Jung K, Kim KJ, Choi HR (2008) A self-sensing dielectric elastomer actuator. Sens Actuators A: Phys 143(2):343–351. doi:10.1016/j.sna.2007.10.076
15. Keplinger C, Kaltenbrunner M, Arnold N, Bauer S (2008) Capacitive extensometry for transient strain analysis of dielectric elastomer actuators. Appl Phy Lett 92(19):192903-1–192903-3. doi:10.1063/1.2929383

Appendices

The following appendices are reprints of earlier publications in the field of electroactivity that were particularly illuminating. This is an incomplete list, as there are many excellent articles and reviews. Prof. Tanaka's mathematical modeling of polyacrylamide gels was helpful with the fundamental understanding of electroactive behavior. Included is a personal favorite by Pelrine, Kornbluh, Pei, and Joseph. Enjoy!

Appendix A: T Tanaka, I Nishio, ST Sun, S Ueno-Nishio, Massachusetts Institute of Technology (1982) Collapse of Gels in an Electric Field. Science 218: 467–469. doi:0036-8075/82/1029-0467, reprinted with permission.

Appendix B: R Pelrine, R Kornbluh, Q Pei, J Joseph, SRI International (2000) High-Speed Electrically Actuated Elastomers with Strain Greater than 100%. Science 287: 836–839. doi:0036-8075/00/0204-0836, reprinted with permission.

L. Rasmussen (ed.), *Electroactivity in Polymeric Materials*,
DOI: 10.1007/978-1-4614-0878-9,

© Springer Science+Business Media New York 2012

Appendix A

T Tanaka, I Nishio, ST Sun, S Ueno-Nishio (1982) Collapse of Gels in an Electric Field. Science 218: 467–469. doi:0036-8075/82/1029-0467. Reprinted with permission.

Reports

Collapse of Gels in an Electric Field

Abstract An infinitesimal change in electric potential across a polyelectrolyte gel produces a discrete, reversible volume change. The volume of the collapsed gel can be several hundred times smaller than that of the swollen gel.

Partially hydrolyzed acrylamide gels in a solvent, such as an acetone-water mixture, undergo discrete and reversible volume transitions upon infinitesimal changes in temperature, solvent composition, pH, or concentration of an added salt [1–3]. The degree of volume change varies with the degree of ionization of the gel and may be as large as 500-fold. The volume change is understood as a phase transition of the system consisting of the charged polymer network, counterions, and fluid [4, 5]. The phase transition is a manifestation of competition among three forces on the gel: the positive osmotic pressure of counterions, the negative pressure due to polymer-polymer affinity, and the rubber elasticity of the polymer network [3]. The balance of these forces varies with changes in temperature or solvent properties.

We demonstrate here that the phase transition is also induced by the application of an electric field across the gel. The electric forces on the charged sites of the network produce a stress gradient along the electric field lines in the gel. There exists a critical stress below which the gel is swollen and above which the gel

L. Rasmussen (ed.), *Electroactivity in Polymeric Materials*,
DOI: 10.1007/978-1-4614-0878-9,
© Springer Science+Business Media New York 2012

collapses. The volume change at the transition is either discrete or continuous, depending on the degree of ionization of the gel and on the solvent composition.

The polyacrylamide gels were prepared by free-radical polymerization [3]. Acrylamide, the linear constituent; N,N′-methylenebisacrylamide, the tetrafunctional cross-linking constituent; and ammonium persulfate and N,N,N′,N′-tetramethylethylenediamine (TEMED), the initiators, were dissolved in water. Micropipettes with a well-defined diameter (1.4 mm) were immersed in this solution. Within 5 min, the solution gelled. After an hour, the gels were removed from the micropipettes and immersed in water to wash away residual monomers. The gels then underwent hydrolysis in a 1.2% solution of TEMED (pH 12) for more than a month. Approximately 20% of the acrylamide groups were converted to acrylic acid groups, some of which were ionized in water,

$$-\mathrm{CONH_2} \rightarrow -\mathrm{COOH} \rightleftharpoons -\mathrm{COO^-} + \mathrm{H^+} \tag{A.1}$$

(In experiments where the composition of the acetone-water mixture is varied, the volume of these gels can change by a factor of 300 with an infinitesimal change in the solvent composition). After hydrolysis, the gels were immersed in a 50% acetone-water mixture. Once equilibrium was reached, the gels were cut into segments 3 cm long. The diameter of each segment was approximately 4 mm.

The cylindrical gel was placed between two platinum electrodes (Fig. A.1a). The voltage applied across the electrodes ranged from 0 to 5 V (Fig. A.1b, c). The electric field produces not only a force on the H^+, causing a stationary current in the gel, but also a force on the negatively charged acrylic acid groups in the polymer network, pulling the gel toward the positive electrode. This latter action creates a uniaxial stress along the gel axis, maximum at the positive electrode and minimum at the negative electrode. The stress gradient deforms the gel. The new shape is in equilibrium within a day. For potentials less than 1 V, there is no drastic change in the gel shape. At 1.25 V, 20% of the gel adjacent to the positive electrode collapses 200-fold in volume. The other 80% of the gel remains swollen. With an increase in voltage, more of the gel collapses. Above 2.15 V, the entire gel collapses. This phenomenon is reversible—when the electric field is removed, the gel assumes the original swollen shape. If the 50% acetone-water mixture is replaced by pure water, the gel diameter changes continuously along the gel axis when the electric field is applied.

The phenomenon can be explained in terms of a mean field theory, initially formulated for gels by Flory and Huggins [6, 7]. Consider a network consisting of ν polymers with their ends cross-linked. Each polymer consists of N_o freely jointed segments, of which f segments are ionized. When formed, the network is cylindrical with length L_o, diameter D_o, and concentration ϕ_o and is in the special condition of no interaction between polymer segments. This special condition is taken as a reference state from which expansion or contraction is measured. A uniform electric field E is applied along the gel axis. One end of the gel is fixed to an electrode to avoid the translational motion of the entire gel along the electric field. Depending on the direction of the electric field, the gel is either compressed or stretched along its axis as a result of the presence of a charged site on the

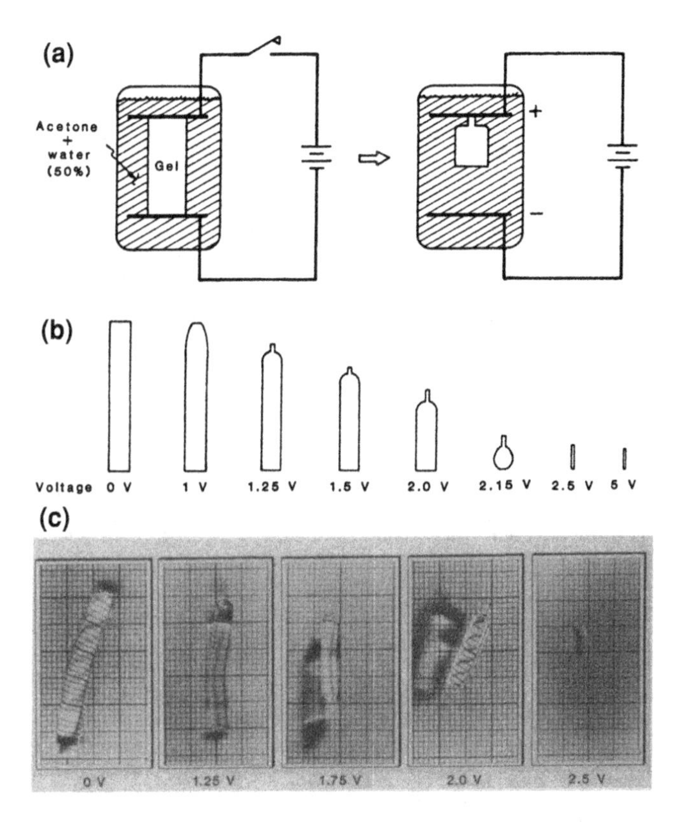

Fig. A.1 Collapse of gel in electric field

network. Consider a thin disk in the gel of thickness ΔZ_o before application of the electric field and located at a distance Z from the free end of the gel (Fig. A.2). Under the effect of E, the thickness changes to $\beta \Delta Z_o$ and the diameter to αD_o. The aim of this analysis is to find the combinations (α, β) that minimize the free energy F of the disk. The free energy includes a term associated with the deformation of the gel F_g and a term for the work done against the electric field F_e,

$$F = F_g + F_e \tag{A.2}$$

For F_g we use the Flory-Huggins formula [7],

$$F_g = \nu kTN_0 \left\{ \frac{1-\phi}{\phi} \ln(1-\phi) + \phi \left[\frac{\Delta F}{kT} \right] \frac{1}{2} \left[2\alpha^2 + \beta^2 - 3 - (2f+1) \ln\left(\alpha^2 \beta\right) \right] \right\} \frac{\Delta Z_0}{L_0} \tag{A.3}$$

where T is the absolute temperature, k is the Boltzmann constant, ϕ is the volume fraction of the polymer network, and ΔF is the free energy decrease associated

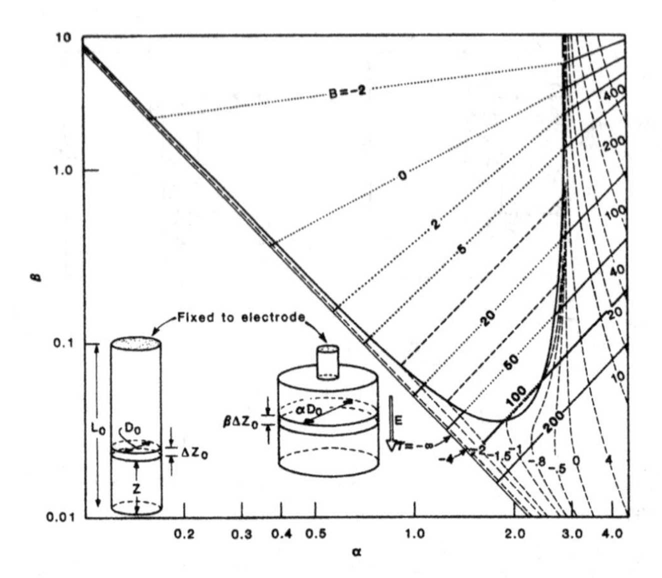

Fig. A.2 Phase diagram of a gel disk under an electric field. Each point in the diagram represents values of the radial expansion α and the axial expansion β which minimize the free energy of the disk for a given combination of the reduced temperature τ and reduced electric potential B calculated from Eqs. A.5 and A.6. The *solid lines* represent constant B for various values of τ, while the *broken lines* represent constant τ for various values of B (B was varied in the experiment). The domain surrounded by the *parabola-like boundary* represents unstable states. The *dotted* and *dashed lines* inside the unstable region connect two stable states with the same free energy. The gel state can change between these two states with an infinitesimal change in B (the *dashed line*) or τ (the *dotted line*)

with a contact between two polymer segments (ΔF varies with the solvent composition). The free energy needed to expand the network against the electric potential is given by

$$F_e = vfeE(Z/L_o)(\beta - 1)\Delta Z_o \equiv BvkT(\Delta Z_o/L_o)(\beta - 1) \qquad \text{(A.4)}$$

where e is the electron charge and $\beta \equiv feE_Z/kT$ is the reduced electric potential.

Minimization of the total free energy yields two equations,

$$\alpha^2 = \beta^2 + B\beta \qquad \text{(A.5)}$$

and

$$\frac{N_0}{\phi_0}\left[\ln(1-\phi) + \phi + (1-\tau)\phi^2/2\right] - (f + 1/2)(\alpha^2\beta) + 1/\beta = 0 \qquad \text{(A.6)}$$

where $\tau \equiv 1{-}2\Delta F/kT$ is the reduced temperature. Equation A.5 shows that the anisotropy of deformation of the disk is uniquely determined by B. If B is positive, the disk is compressed more in the axial direction than radially ($\alpha > B$). The reverse is true if B is negative. For certain values of τ and B, Eqs. A.5 and A.6 have two solutions corresponding to two minima of free energy. In this case, the smaller free energy minimum represents the stable state. A discrete volume transition occurs from minimum to the other when the values of two minima coincide. These results are summarized in Fig. A.2, where the stable states (α, β) are plotted at various combinations of τ and B. Figure A.2 shows that the discrete transition can occur when τ is changed for a fixed B or when B is varied for a fixed τ. The magnitude of the discrete transition is greater for lower τ and B. At larger values of these parameters, the transition becomes continuous. This analysis accounts for both the discrete transition observed in a 50% acetone-water mixture and the continuous transition in pure water. The former corresponds to a low τ value, whereas the latter corresponds to a high τ value. Although only the electric force was considered in describing the phase transition, other factors may play a role. The effects of inhomogeneous distributions of ions, currents, and solvent composition within the gel and electrochemical reactions occurring at the electrodes need to be considered for a complete understanding of the phenomena.

The discrete volume transitions of the gel induced by an electric field can be used to make switches, memories, and mechanochemical transducers. For example, ionic gels controlled by coordinated signals from a microcomputer may be used for an artificial muscle. It may also be possible to store two- or three-dimensional images by using the local collapse and swelling of the gel.

Toyoichi Tanaka
Izumi Nishio
Shao-Tang Sun
Shizue Ueno-Nishio

Department of Physics and Center
for Materials Science and Engineering
Massachusetts Institute of Technology
Cambridge, MA 02139, USA

References

1. Tanaka T (1978) Sci Am 244:124 (Jan 1981)
2. Tanaka T (1978) Phys Rev Lett 40:820
3. Fillmore D, Nishio I, Sun S-T, Swislow G, Shah A (1980) ibid 45:1936
4. Tanaka T, Ishiwata S, Ishimoto C (1977) ibid 38:771
5. Tanaka T (1978) Phys Rev A 17:763
6. Dusek K, Patterson D (1968) J Polym Sci Part A-2 6:1209
7. Flory PJ (1953) Principles in polymer chemistry. Corndll University Press, Ithaca
8. We thank Swislow G and Ohmine I for their critical reading of the manuscript. This work hasbeen supported by the Office of Naval Research under grant N00014-80-C-0500. 4 Jan 1982; revised I June 1982

Appendix B

R Pelrine, R Kornbluh, Q Pei, J Joseph (2000) High-Speed Electrically Actuated Elastomers with Strain Greater Than 100%. Science 287: 836–839. doi: 10.1126/science.287.5454.836 Reprinted with permission.

High-Speed Electrically Actuated Elastomers With Strain Greater Than 100%

Ron Pelrine,* Roy Kornbluh, Qibing Pei, Jose Joseph
SRI International, 333 Ravenswood Avenue, Menlo Park, CA 94025, USA.
*To whom correspondence should be addressed. Email: pelrine@erg.sri.com

Electrical actuators were made from films of dielectric elastomers (such as silicones) coated on both sides with compliant electrode material. When voltage was applied, the resulting electrostatic forces compressed the film in thickness and expanded it in area, producing strains up to 30–40%. It is now shown that prestraining the film further improves the performance of these devices. Actuated strains up to 117% were demonstrated with silicone elastomers, and up to 215% with acrylic elastomers using biaxially and uniaxially prestrained films. The strain, pressure, and response time of silicone exceeded those of natural muscle; specific energy densities greatly exceeded those of other field-actuated materials. Because the actuation mechanism is faster than in other high-strain electroactive polymers, this technology may be suitable for diverse applications.

New high-performance actuator materials capable of converting electrical energy to mechanical energy are needed for a wide range of demanding applications, such as mini- and microrobots, micro air vehicles, disk drives, flatpanelloudspeakers, and prosthetic devices. Many types of candidate materials are under investigation, including single-crystal piezoelectric ceramics [1] and carbon nanotubes [2]. Electroactive polymers are of particular interest because of the low cost of materials and the ability of polymers to be tailored to particular applications. Within the general category of electroactive polymers, many different

L. Rasmussen (ed.), *Electroactivity in Polymeric Materials*,
DOI: 10.1007/978-1-4614-0878-9,

© Springer Science+Business Media New York 2012

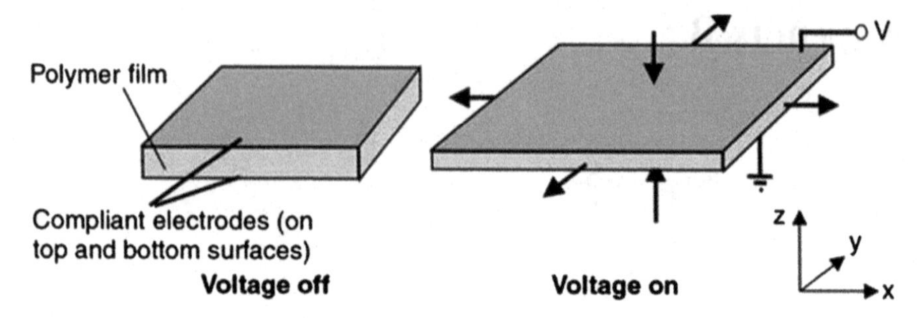

Fig. B.1 The dielectric elastomers actuate by means of electrostatic forces applied via compliant electrodes on the elastomer film

types are under investigation, including electrostrictive polymers [3, 4], piezoelectric polymers [5], and electrochemically actuated conducting polymers and gels [6–11]. Most electroactive polymers excel in some measures of performance (such as energy density or strain) but are unsatisfactory in others (such as efficiency and speed of response).

It has been well known for many years that the electric field pressure from free charges on the surface of all insulating materials induces stresses (Maxwell's stress) that strain the material. Zhenyi et al. [4] showed that a largely noncrystalline polymer (polyurethane) could produce actuated strains of 3–4% using metal electrodes such as 20 nm-thick gold; they estimated that 10% of their observed strain response was due to Maxwell stress. More recently, it has been suggested that Maxwell stress by itself can produce powerful electroactive responses in certain elastomers [12, 13]. This mechanism of actuation (Fig. B.1) distinguishes dielectric elastomers from most electrostrictive polymers previously reported. A dielectric elastomer film, typically 10–200 μm thick, is coated on each side with a compliant electrode material (e.g., carbon-impregnated grease). When a voltage is applied across the two electrodes, the electrostatic forces compress and stretch the film. Compression of the film thickness brings opposite charges closer together, whereas planar stretching of the film spreads out or separates similar charges. Both changes convert electrical energy to mechanical energy and provide the actuation mechanism.

The actuation mechanism illustrated in Fig. B.1 was previously shown to have high actuation pressures (0.1–2 MPa), fast response times (<1 ms), and potentially high efficiencies (>80–90%) [12]. A variety of elastomer materials have been investigated, including silicones such as NuSil Technology's CF19-2186 and Dow Corning's HS3. Peak strains of 32% (CF19-2186) and 41% (HS3) were demonstrated, with specific elastic energy densities up to 0.15 J/g for CF19-2186 [12].

We have now conducted experiments that demonstrate extraordinarily high strains, five to six times those previously reported, with higher pressures (up to 7 MPa) and energy densities about 23 times those described earlier. The improvement is due to the identification of a new dielectric actuator material

(3M's VHB 4910 acrylic) as well as the application of high prestrain in one planar direction, which enhances electrical breakdown strength and causes the material to actuate primarily in the low-prestrain planar direction. We also present data on applying higher prestrains to improve the performance of previously described silicones. Higher strains and actuation pressures can potentially be exploited to improve a wide range of existing actuator devices (e.g., pumps, motors, robot actuators, generators, and flat-panel loudspeakers) as well as enable new applications (e.g., small flapping-wing vehicles, lifelike prosthetics, noise suppression devices, and biologically inspired robots).

The compression and stretching modes of actuation are mechanically coupled for most elastomers because, at the stresses of interest, the elastomer volume is essentially fixed (the bulk modulus is much higher than the modulus of elasticity Y). We can use our electrostatic model to show that the effective compressive stress, p, compressing the film in thickness [13] is

$$p = \varepsilon\varepsilon_0 E^2 \tag{B.1}$$

where ε is the relative dielectric constant of the material, ε_0 is the permittivity of free space (8.85×10^{-12} F/m), and E is the electric field (volts per meter). The effective compressive stress in Eq. B.1 is twice the stress normally calculated for two rigid, charged capacitor plates, because in an elastomer the planar stretching is coupled to the thickness compression. We refer to the stress in Eq. B.1 as an effective stress because, strictly speaking, it is the result of both compressive stress acting in the thickness direction and tensile stresses acting in the planar directions. The compressive and tensile stresses are mechanically equivalent in a thin film to a single compressive stress acting in the thickness direction according to Eq. B.1.

For low strains (e.g., <20%), the thickness strain s_z can be approximated by

$$s_z = -p/Y = -\varepsilon\varepsilon_0 E^2/Y \tag{B.2}$$

For strains greater than about 20%, Eq. B.2 is unsatisfactory because Y generally depends on the strain itself. The high actuated strains we observed require the modification of other conventional actuator material constitutive relations as well, even with an assumption of constant modulus. For example, the elastic strain energy density in an actuator material, u_e (a common parameter for comparing the output capabilities of actuator materials), is typically expressed as $u_e = \frac{1}{2}ps_z = \frac{1}{2}Ys_z^2$, but this formula assumes low strains. For high strains, the planar area over which the compression acts increases substantially as the material is compressed [12]. For high-strain, nonlinear materials, where the compressive stress is known, a more useful measure of performance might be the electromechanical energy density e, which we define as the amount of electrical energy converted to mechanical energy per unit volume of material for one cycle. The electromechanical energy density can be written as

$$e = -p\ln(1 + s_z) \tag{B.3}$$

[14], where p is the constant compressive stress. If we substitute $p = Ys_z$ on the right side of Eq. B.3 and expand the logarithm for small s_z, it follows that $\frac{1}{2}e = u_e$, thus providing a valid comparison between the high-strain, nonlinear materials discussed here and conventional low-strain energy density formulas for materials such as piezoelectrics.

We tested many types of polymeric elastomer films. Here we focus on three promising types: Dow Corning HS3 silicone, NuSil CF19-2186 silicone, and the 3M VHB 4910 acrylic adhesive system [15]. As noted above, results for the silicone films were reported in earlier publications, but new high-prestrain results using these polymers are reported here.

Strain measurements were made with elastomer films stretched on a rigid frame. Compliant electrodes were stenciled with conductive carbon grease (Chemtronics Circuit Works CW7200) on the top and bottom of the films. The active, electroded portion of the stretched film was small relative to the film's total area. Thus, the inactive portions of the film acted as a spring force on the boundaries of the active regions. When a voltage difference was applied between the top and bottom electrodes, the active region expanded while the inactive region contracted. Removing the applied voltages caused the reverse change. A digital video optical system was used to measure the actuated strain. Measurements were taken about 1 s after application of the voltage. The stretched film technique for measuring strains introduces some boundary constraints from the inactive portion of the film, but it circumvents the difficulty of trying to achieve free-boundary conditions with a soft flexible material [12].

For an elastomer, the absolute strain under actuation depends on the prestrain. A more useful quantity is the relative strain under actuation:

$$\frac{(\text{actuated length}) - (\text{unactuated length})}{(\text{unactuated length})}$$

The relative strain equals the absolute strain if there is zero prestrain in the film. The relative area strain is defined similarly, with the active planar area replacing length in the above expression.

Two types of strain tests were performed, circular (biaxial) and linear (uniaxial). In the circular tests, a small circular active region (5 mm in diameter) was used to decrease the likelihood of a fabrication defect causing an abnormally low breakdown voltage. The film was stretched uniformly on the frame, and the circle expanded in area when a voltage was applied (Fig. B.2). The expansion of the circle is equal in both x and y planar directions because there is no preferred planar direction for the film. By contrast, the linear strain tests used a high prestrain in one planar direction and little or no prestrain in the other planar direction. High prestrain effectively stiffens the film in the high-prestrain planar direction, which causes the film to actuate primarily in the softer, low-prestrain planar direction and in thickness. Figure B.3 shows a linear strain test. The relative strain was measured in the central region of the elongated (black) active area, away from the edge constraints.

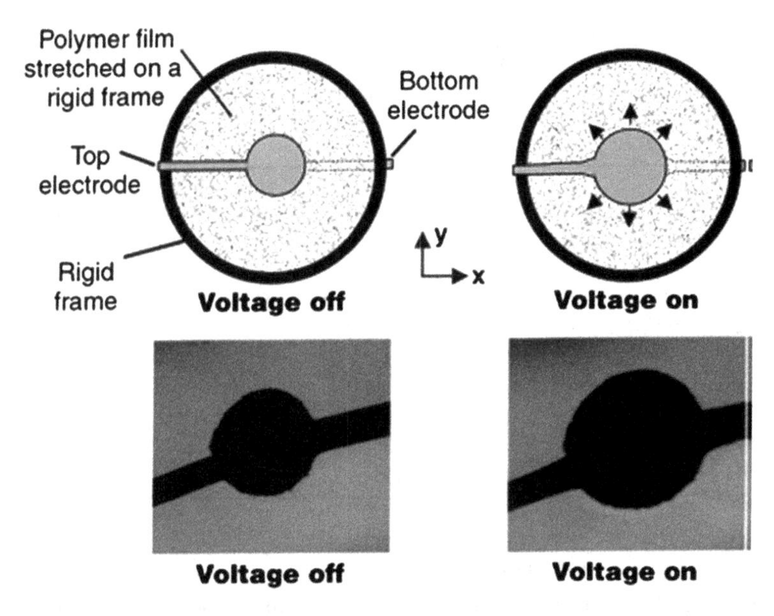

Fig. B.2 The circular strain test measures the expansion of an actuated circle on a larger stretched film. The photo shows 68% area expansion during actuation of a silicone film

The circular test results for three elastomers under different conditions of prestrain are given in Table B.1. The peak relative area strain was measured directly, and the relative thickness strain was calculated from the constant volume constraint. The breakdown field was calculated from the known voltage and the measured film thickness (corrected for the given relative thickness strain). No attempt was made to minimize voltage with these relatively thick films, and voltages were typically 4–6 kV. Thinner films generally yield lower but comparable performance at lower voltage. For example, preliminary measurements showed 104% relative area strain at 980 V using a thinner acrylic film. The electromechanical energy density e was estimated from the peak field strength (Eq. B.1) and the relative thickness strain. The value ½e is listed in Table B.1 for convenient comparison to conventional elastic energy densities available for other actuator materials.

As indicated by the values, the VHB 4910 acrylic elastomer gave the highest performance in terms of strain and actuation pressure. Extensive lifetime tests have not been made, but acrylic films have been operated continuously for several hours at the 100% relative area strain level with no apparent degradation in relative strain performance. However, the acrylic elastomer has relatively high viscoelastic losses that limit its half-strain bandwidth (the frequency at which the strain is one-half of the 1-Hz response) to about 30–40 Hz in the circular strain test. By comparison, HS3 silicone has been used for prototype loudspeakers at frequencies as high as 2–20 kHz [16, 17]. The actuation of CF19-2186 silicone, albeit at lower strains and fields than reported here, has been measured directly via laser reflections with full

Table B.1 Circular and linear strain test results

Material	Prestrain (*x*, *y*) (%)	Actuated relative thickness strain (%)	Actuated relative area strain (%)	Field strength (MV/m)	Effective compressive stress (MPa)	Estimated 1/2*e* (MJ/m^3)
Circular strain						
HS3 silicone	(68, 68)	48	93	110	0.3	0.098
	(14, 14)	41	69	72	0.13	0.034
CF19-2186 silicone	(45, 45)	39	64	350	3.0	0.75
	(15, 15)	25	33	160	0.6	0.091
VHB 4910 acrylic	(300, 300)	61	158	412	7.2	3.4
	(15, 15)	29	40	55	0.13	0.022
Linear strain						
HS3	(280, 0)	54	117	128	0.4	0.16
CF19-2186	(100, 0)	39	63	181	0.8	0.2
VHB 4910	(540, 75)	68	215	239	2.4	1.36

strain response up to 170 Hz (resonance effects prevented measurement at higher speeds) [12]. The only apparent fundamental limits on actuation speed are the viscoelastic losses, the speed of sound in the material, and the time to charge the capacitance of the film (electrical response time).

The strains in the linear strain test can be quite large, up to 215% for the VHB 4910 acrylic adhesive (Table B.1). The VHB 4910 acrylic elastomer, when undergoing ~160% strain in a linear strain test, exhibited buckling (the vertical wrinkles in Fig. B.3d) that was not seen in properly stretched silicone films. Buckling indicates that the film is no longer in tension in the horizontal direction during actuation, and that the overall relative thickness strain is greater than indicated by measurements of the electrode boundaries. That is, the relative strain numbers for VHB 4910 in Table B.1 may be undervalued.

The dielectric elastomer films presented here appear promising as actuator materials because their overall performance can be good. The available literature indicates that the actuated strains of silicone are greater than for any known high-speed electrically actuated material (that is, a bandwidth above 100 Hz). Silicone elastomers also have other desirable material properties such as good actuation pressures and high theoretical efficiencies (80–90%) because of the elastomers' low viscoelastic losses and low electrical leakage [12].

The VHB 4910 acrylic adhesive appears to be a highly energetic material. The energy density of the acrylic adhesive is three times that reported for single-crystal lead–zinc niobate/lead titanate (PZN-PT) piezoelectric (about 1 MJ/m3) [1], itself an energetic new material with performance much greater than that of conventional piezoelectrics. The density of both the silicones and the acrylic adhesive is approximately that of water and about one-seventh that of ceramic piezoelectric materials. Hence, the energy density of the acrylic adhesive on a per-weight basis (the specific energy density) is about 21 times that of single-crystal

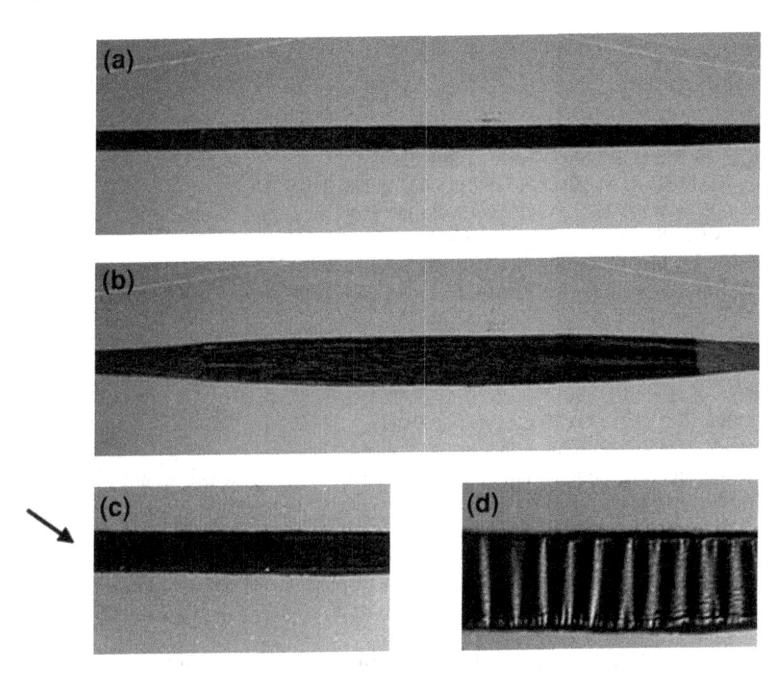

Fig. B.3 (**a** and **b**) Linear strain test of HS3 silicone film with a high horizontal prestrain for the field off (**a**) and on (**b**) with a field of 128 V/μm; 117% relative strain was observed in the central region of (**b**). (**c** and **d**) Activation of acrylic elastomers, producing about 160% relative strain, for the field off (**c**) and on (**d**); the dark area in (**c**) indicates the active region

piezoelectrics and more than two orders of magnitude greater than that of most commercial actuator materials.

Potential applications for dielectric elastomer actuators include robotics, artificial muscle, loudspeakers, solid-state linear actuators, and any application for which high performance actuation is needed. A variety of actuator devices have been made with the silicone elastomers, including rolled actuators, tube actuators, unimorphs, bimorphs, and diaphragm actuators [12, 18, 19]. Their performance is promising, but most of this work did not exploit the benefits of high prestrain or the new acrylic material. We have built an actuator using 2.6 g of stretched acrylic film that demonstrated a force of 29 N and displacement of 0.035 m, a high mechanical output for such a small film mass. The very high strains recently achieved suggest novel applications for shape-changing devices, and the specific energy density of the acrylic adhesive is so high that, if it could be realized in a practical device, it could replace hydraulic systems at a fraction of their weight and complexity. However, practical applications require that a number of other issues be addressed, such as high-voltage, high-efficiency driver circuits, fault-tolerant electrodes, long-term reliability, environmental tolerances, and optimal actuator designs.

References

1. Park S, Shrout T (1997) J Appl Phys 82:1804
2. Baughman RH et al (1999) Science 284:1340
3. Zhang QM, Bharti V, Zhao X (1998) Science 280:2101
4. Zhenyi M et al (1994) J Polym Sci B Polym Phys 32:2721
5. Furukawa T, Seo N (1990) Jpn J Appl Phys 29:675
6. Smela E, Inganas O, Lundstrom I (1995) Science 268:1735
7. Smela E, Gadegaard N (1999) Adv Mater 11:953
8. Otero T et al (1999) Proc SPIE 3669:98
9. Schreyer HB et al (1999) Proc SPIE 3669:192
10. Oguro K et al (1992) J Micromachine Soc 5:27
11. Tamagawa H et al (1999) Proc SPIE 3669:254
12. Kornbluh R et al (1999) Proc SPIE 3669:149
13. Pelrine R et al (1998) Sens Actuators A Phys 64:77
14. For a constant-volume material, $(1 + s_x)(1 + s_y)(1 +1\ s_z) = 1$, where s_x and sy are the length and width strains. As the film is squeezed, the area of compression $A(s_z)$ can be expressed as $A(s_z) = xy = x_o(1 + s_x)y_o(1 + s_y) = (x_oy_oz_o)/[z_o(1 + s_z)]$, where x_o, y_o, and z_o are the initial length, width, and thickness of the active area of the film. The energy density converted to mechanical work is then the integral, over the displacement, of the compressive stress times the area of compression divided by the volume:

$$e = -(1/x_o\ y_oz_o) \int pA(s_z)dz = -\int p[1/(1 + s_z)]ds_z = -p \ln(1 + s_z)$$

where p is the assumed constant compressive stress. The minus sign is introduced because we are defining p as a positive number for compression (dz is negative over the integration). The assumption of constant p depends on the electronic driving circuitry, which ideally would adjust the applied voltage according to the varying thickness to hold the electric field constant. It can be shown that with a nonideal, constant-voltage drive, the term $\ln(1 + s_z)$ would be replaced by $-(s_z + 0.5s_z^2)$. However, because the present focus is on the fundamental material performance rather than electronic performance, we make the simplest physical assumption that p is constant.
15. The silicone films are based on a polydimethyl siloxane backbone. They were diluted in naphtha solvent, spin-coated, cured, and released. The HS3 silicone was centrifuged to remove pigment particles before spin coating. VHB 4910 is available in film form with a removable liner backing. The acrylic elastomer is made of mixtures of aliphatic acrylate photocured during film processing. Its elasticity results from the combination of the soft, branched aliphatic groups and the light cross-linking of the acrylic polymer chains. The zero-strain thicknesses of the materials were typically 225 μm for HS3, 50 μm for CF19-2186, and 1000 μm for the VHB 4910 acrylic. The relative dielectric constant at 1 kHz is 2.8 for the two silicones and was measured at 4.8 + 0.5 for the VHB 4910 acrylic.
16. Heydt R et al (1998) J Sound Vibr 215:297
17. R. Heydt et al J Acoust Soc Am (in press)
18. Kornbluh R et al (1995) Proceedings of the third IASTED international conference on robotics and manufacturing, Cancun, Mexico, 14 to 16 June 1995. ACTA Press, Calgary, Alberta, pp 1–6

19. Kornbluh R et al (1998) Proceedings of the 1998 IEEE international conference on robotics and automation, Leuven, Belgium, May 1998. IEEE Press, Piscataway, NJ, pp 2147–2154
20. Much of this work was performed under the management of the Micromachine Center at the Industrial Science and Technology Frontier Program, Research and Development of Micromachine Technology of MITI (Japan), supported by the New Energy and Industrial Technology Development Organization. 7 September 1999; accepted 6 December 1999

Index

L. Rasmussen (ed.), *Electroactivity in Polymeric Materials*,
DOI: 10.1007/978-1-4614-0878-9,
© Springer Science+Business Media New York 2012